BENNETT BOOKS
Science of Mind Series

Bennett Books' *Science of Mind Series* presents works that facilitate new ways of seeing and understanding. These studies ask how the underlying wholeness and harmony of the world can be explored and expressed in languages appropriate for our current human situation and how, in turn, the resulting awareness can foster both individual self-development and practical real-world change.

○

This is the first book of the *Science of Mind Series*
and is printed on 60 lb. acid-free paper.

In Memory of

Elizabeth Bennett
and
Michael Franklin

Elementary Systematics
A Tool for Understanding Wholes

J.G. Bennett, 1953

Elementary Systematics
A Tool for Understanding Wholes

J.G. Bennett

•

Edited by David Seamon

BENNETT BOOKS
Santa Fe, New Mexico

Edited by David Seamon from six lectures given by Bennett, October
through December, 1963, at the Institute for the Comparative Study
of History, Philosophy, and the Sciences, Coombe Springs,
Kingston-on-Thames, London, England.

BENNETT BOOKS
P.O. Box 1553
Santa Fe, New Mexico 87504

Book Design: AM Services
Cover Illustration: Robert Armon

Library of Congress Cataloging-in-Publication Data
Bennett, John G. (John Godolphin), 1897-1974.
    Elementary Systematics: A tool for understanding wholes / J.G.
  Bennett ; edited by David Seamon.
      p. cm. -- (Science of mind series)
    "Edited . . . from six lectures given . . . October-December 1963 at
the Institute for the Comparative Study of History, Philosophy, and the
Sciences, Coombe Springs, Kingston-on-Thames, London, England"
–T.p. verso.
    Includes bibliographical references.
    ISBN 0-9621901-7-9 (pbk. : alk. paper)
    1. Symbolism of numbers.   2. Whole and parts (Philosophy)
I. Seamon, David.   II. Title.   III. Series.
BP610.B45 1992
133.3'35--dc20   92-24792
                                         CIP

○

About the Cover

The cover illustration, by Robert Armon, is a pattern based on traditional Islamic
geometry and represents unity in diversity, a philosophical system of parts that work
together to make a whole. Systematics is one such philosophical system.

Armon is a registered architect in Colorado and New Mexico with offices in Denver
since 1975. Armon has worked with sacred geometry and architecture for 25 years.
He studied with Dr. Keith Critchlow, FRCA, and the School of Sacred Arts .

Contents

Foreword

THESE SIX LECTURES, delivered in England more than a quarter of a century ago, were among the earliest of J. G. Bennett's presentations of systematics. They are characterized by the author's customary clarity and his simplicity in the presentation of complex ideas. His examples of the working of systems range from writing with chalk on a blackboard to the revolution of the planets.

Several more scholarly and "difficult" demonstrations of systematics have been made over the years, by Bennett himself and by others, usually to specialist audiences. Now, the admirable simplicity in the delivery of the six lectures makes the approach available for the first time to any seeker, with or without any special training.

I am personally very pleased that these lectures will now be accessible to the general reading public, and I would like to express my grateful thanks to David Seamon, whose meticulous and sympathetic editing has made this presentation an actuality.

Elizabeth Bennett
June, 1990
England

Editor's Introduction

JOHN G. BENNETT (1897–1974) was a British scientist, mathematician and philosopher who integrated scientific research with studies of Asiatic languages and religions. *The Dramatic Universe,* his influential masterwork, was published in four volumes between 1956 and 1966.[1] Throughout his life, Bennett travelled widely and met many little-known but important spiritual leaders. In the early 1920s, he was introduced to George Gurdjieff (1877–1949) and Peter Ouspensky (1878–1947), two philosophers and spiritual teachers who became central guiding forces in his life.

One important theme for both Gurdjieff and Ouspensky was the symbolic significance of numbers—a topic important in many religious traditions, both East and West. As a mathematician, Bennett became interested in reconciling the quantitative, materialist approach to number in modern western science with the qualitative, intuitive approach of the world's spiritual traditions. In 1946, he founded the Institute for the Comparative Study of History, Philosophy and the Sciences, in Coombe Springs, Kingston, a suburb of London. One aim of the Institute was to sponsor research that explored the qualitative significance of number. In time, Bennett and his colleagues developed an approach they called *systematics*. This book is an introduction to that work.

What is systematics? In broadest terms, it is a conceptual tool that helps one to find pattern, order, and meaningful progression in the midst of life's complexity and confusion. Systematics assumes that, beneath the world's diversity and continual change, there is an underlying coherence that provides a vehicle for understanding, both practical and theoretical. Whatever the particular subject or task that one is involved with, systematics argues that there are certain recognizable patterns that help one to move through the subject's complexity and subtlety and to see the subject as it is essentially.

For systematics, the basis of these recognizable patterns is *the qualitative significance of number.* Systematics assumes that the world has an underlying coherence that can be described through the experienced qualities of number, thus the qualitative meaning of *one-ness* helps to define the particular whole in which one is interested, while the qualitative meaning of *two-ness* helps to define the various differences, polarities, and complementarities present in the whole. Yet again, *three-ness* helps to define relationship and reconciliation, while *four-ness* helps to define activity; and *five-ness*, significance and potential.

Bennett uses the word *system* to designate the underlying pattern that a specific number represents. Further, by using the Greek word for the particular number followed by the suffix *-ad,* he gives each system a name. Thus the *monad* represents one-ness; the *dyad* two-ness; the *triad,* three-ness; the *tetrad,* four-ness; the *pentad,* five-ness; and so forth. Bennett argues that each of these systems can be applied to the particular thing in which the student is interested. In this way, he or she will gain a more comprehensive understanding of the thing and be better able to appreciate and work with it. In this book, Bennett introduces the systems from monad to pentad.

One way to describe systematics is to distinguish between two contrasting approaches to knowledge—analysis and synthesis. In analysis, the student simplifies by breaking the thing into parts and applying some method of examination. The aim is to explain how the thing works and to formulate practical action or theoretical pattern. Analysis is an integral part of modern western science, which has been immensely successful in broadening our practical knowledge, but less successful in regrouping all the parts to make again the original whole.

Another way to understand the world is through synthesis, which accepts the complexity of the situation, but hopes to find underlying patterns that will allow the student to see the situation holistically in a clearer, simpler way. In the late twentieth century, we realize that nothing is either simple or self-evident. The key question, in terms of synthesis, is whether, in the midst of the world's complexity, we can find some order other than the patterns of analysis, which are often imposed and reductive.

Drawing on a symbolic interpretation of numbers, systematics is one tool for synthesis. Through underlying patterns arising from the qualitative significance of numbers, systematics works to stay in contact with the whole rather than to break it into parts that are then studied piecemeal. Throughout his writings, Bennett emphasizes that analysis reigns in the twentieth century, particularly because of the practical, technological success of materialist science. He also emphasizes, however, that the human world is not, ultimately, a machine and cannot be understood or repaired in the same way that a machine can. He, therefore, identifies a need to come closer to the world and to see it in a more comprehensive and accurate way. He deeply believes that systematics offers one conceptual means to answer this need.

The present book is an edited transcription of six lectures that Bennett gave in late 1963 at his Coombe Springs Institute.[2] The first lecture discussed differences between knowledge and understanding, while the remaining five lectures described the systems from monad to pentad. These lectures were entirely spontaneous, and Bennett appears not to have reworked or edited the lectures in written form. They were, however, transcribed to typescript and over the years, circulated among a small group of people interested in systematics.

Several years ago, Saul Kuchinsky, one of the senior students of systematics today, distributed the six lectures to interested individuals in Great Britain and North America. It was then that I first read the lectures and was struck by their clarity and force. The strength of the lectures, however, was also their weakness: the spontaneity and power of Bennett's spoken words were sometimes lost in the fixed form of a written text. Often, there were gaps in meaning, or a point arose for which the reader was unprepared. People with a background in systematics could more or less follow the lectures, but newcomers frequently found them confusing and obscure.

In preparing the six lectures for publication, therefore, my main aim has been *clarity and accessibility*. Toward this end, I have actively worked on the lectures' original text and made three kinds of changes. First, I have frequently made shifts in word and sentence phrasing, since what is conveyed in speaking and what is conveyed in writing are often very much different. This change was required throughout the six lectures, and my aim was to work for smoother flow and more lucid meaning. Second, and less frequently, I have made changes in the order in which sections of the lectures were presented. These changes were required where themes are introduced for which the reader is unprepared. Third, and largely in the chapter on triads, I have written new sections (mostly introductory paragraphs) without which meaning would falter. I have indicated these new sections in the end notes and have based their content on Bennett's other written comments on the particular topic.

The end notes for the lectures are of three types. Occasionally, I felt there was important material in the original lecture that impeded the flow of meaning. I have incorporated this material in notes that begin and end with double quotation marks; these notes present passages that were spoken by Bennett directly. A second group of notes clarifies material with which newcomers may be unfamiliar, and a third group provides clarifications that Bennett made about a particular topic in his other writings.

The book also includes two appendices, the second of which is a list of further reading for individuals who wish to learn more about systematics. The first appendix is a reprint of Bennett's "General Systematics," originally published in the inaugural issue of *Systematics*, the Institute's research journal that was published from 1963 to 1974. Like the six lectures, this article is an introduction to systematics but discusses systems up to the eight-term octad and is more difficult than the six lectures.

Saul Kuchinsky and I decided to include this article because it extends many points that Bennett discussed in the lectures and illustrates a more academic, intellectual presentation of systematics. In editing this article, I have made small changes and additions that include minor phrasing revisions, notes citing works or discussions that Bennett mentions but for which he provides no or incomplete references, and subheadings to make the order of the article more clear. "General Systematics" is more compressed and philosophical than the six lectures, and the reader is advised to study the lectures first before reading the article.

Systematics offers a valuable way for making sense of the world and for discovering a fuller meaning in life. There is a need for a clear introduction to the style and method of systematics, and I hope that *Elementary*

Systematics helps to fulfill this need. Over the years, other people have also believed in the value of systematics, and some of these individuals have helped me greatly in my editing task. I must thank three people directly: first, the late Elizabeth Bennett, for her confidence that I could transform the spoken lectures into clear written form; and, second, the late Michael Franklin, for insisting on the need for this book and for reviewing my first draft of the edited lectures.

Particularly, I am grateful to Saul Kuchinsky who profoundly believes in systematics and has helped to keep the approach alive over the last twenty years. He first suggested this enterprise and nudged me to completion. He also provided important corrections and suggestions that make the book richer and stronger. In many ways, *Elementary Systematics* is as much his book as Bennett's, and if systematics is to endure, it will partially be because of Saul Kuchinsky's persistent and selfless efforts.

David Seamon

NOTES

1. Full references for these volumes, as well as a list of other relevant studies, are provided in appendix B, "Further Reading."

2. In *Systematics*, the journal of Bennett's Institute, there is listed in the Institute Calendar for October, 1963, the following description of the lectures: "Systematics for Beginners, October 11-December 13, alternative Fridays" (vol. 1, no. 2 [1963], p. 183).

Elementary Systematics
A Tool for Understanding Wholes

Systematics, Knowing, and Understanding

SYSTEMATICS is an instrument of understanding. The basis of systematics is the belief that there are principles according to which everything can be described and understood. These principles demonstrate that all structures in the world, be they things, living beings, events or processes, can be understood by simple patterns or *systems*, each of which can be expressed in terms of one characteristic quality, such as universality, complementarity, dynamism, potentiality, or completedness. These qualities, in turn, arise from the experienced significance of number—that is, the qualitative meaning of oneness, twoness, threeness, and so forth.

For example, the first system in systematics relates to oneness, for which the characteristic quality is *wholeness*—the fact that every structure is a totality that can be considered as a whole in spite of a diversity of content. This system is called the *monad* and reflects universality and unity in diversity. The monad is presented in chapter 1 of this book.

If a structure can be studied in terms of oneness and the monad, it can also be studied in terms of twoness, which leads to *differences* and the system of the *dyad*. We could not know or understand anything if we could not recognize differences, which must therefore be a universal property of all structures. We can look at any structure in complementary ways that must be considered together to see and understand the whole.

When we look at a person, for example, we know that he or she has an inner or private life that we call *consciousness* and an outer or public life that we call *behavior*. We cannot understand the person in terms of either consciousness or behavior alone. These inner and outer qualities are complementary, not only for us as observers, but also for the person as he or she is. When these complementary parts are combined, they make up the life and process of a human being. The dyad allows one to understand a structure in terms of differences, polarity, and complementarity. Chapter 2 discusses the dyad.

In this book, we consider systems up to five. Chapter 3 discusses threeness and the *triad*; chapter 4, fourness and the *tetrad*; and chapter 5, fiveness and the *pentad*. Systematics argues that consideration of a structure from the points of view of the different systems provides a practical, comprehensive understanding.

UNDERSTANDING AND KNOWLEDGE

Before presenting the method of systematics in detail, it is important to discuss differences between **understanding** and **knowledge**. The primary aim of systematics is to develop the power of understanding, which differs from knowledge in that it cannot be taught and comes only when one sees for oneself. This book can only help you by providing an opportunity to "see" for yourself. To see, we must be able to recognize what we are looking at, which means that knowledge must come before understanding. We have much knowledge about many subjects, and you may read this book because you wish to understand one particular subject better.

To understand is *to see the way things belong together, and to see why they are together as they are.* Understanding relates to underlying patterns, relationships, and meanings. If each thing of the world were different from everything else, we should have no hope of understanding. We would have to rely upon knowing everything uniquely and piecemeal. Probably animals know things in this fragmented way, whereas understanding is distinctly human. The urge to understand ourselves and the world in which we live is one of the noblest and most powerful human motivations. Often, however, when we ask for the bread of understanding, we are fed with the stone of knowledge.

Experience is a key to understanding. Though it is not necessary that we should have experienced the particular thing that we wish to understand, it is necessary that we should be able to relate it to something that we have experienced. Understanding can be transferred from one situation to another, and this property distinguishes it from knowing. On the other hand, understanding cannot be communicated, whereas knowledge can. However much I may wish to share with another person my understanding of something, I will be completely unable to unless we share some common experience of that particular thing. It is impossible to communicate this understanding in the way we can communicate knowledge.

Knowing is a power that can be communicated but not transferred. Understanding is a power that can be transferred but not communicated. If I know, for example, where and when one catches a train to London, I can pass this travel information on to you. But the knowledge that I have about trains to London does not tell anything about trains from, let us say, Budapest to Prague. That is something that I have to learn separately from someone who knows and can pass the information on to me. Similarly, with anything that one knows, if it really is knowledge, one can communicate it, but knowledge of one situation cannot be transferred to another situation, except in so far as there is something in common between the two situations.

It is just the opposite with understanding, which cannot be communicated but can be transferred from one situation to another. The understanding of one experience can be applied to the understanding of similar experiences, even if they happen in a different place or in a different form. For example, if I understand the structure of language and what language is, then I understand all languages, but I cannot get this understanding across to you unless you have experience in linguistics, as I have. On the other hand, if I know a language, I can teach it to you and thus pass it on. It does not follow, however, that if I know one language, I know all languages.

Understanding can be transferred from one situation to another because there are simple, recurring patterns underlying all things and situations. If we learn to see these patterns in different outward forms, we will also be able to recognize them in new, unfamiliar situations. We are able to transfer the underlying patterns from one situation to another. When I say that these patterns underlie outward forms and are relatively simple, I do not refer to qualities established by seeing and touching, such as "roundness" and "squareness." Instead, these patterns are much more connected with the very nature of things, which in turn means that they are connected with our own human nature.

The property of joining individual human natures with natures other than their own is one reason why understanding is a special power for people. It is said that if we understand ourselves, we understand the world. And if we understand the world, we understand ourselves. This transference from one situation to another is one of the most important properties of understanding—that it can be transferred from inside to outside. If I understand how things happen inwardly, I shall also understand how things happen outwardly. This reciprocity means that my own nature and world of inner experience is constructed according to the same pattern as the world of my outer experience. Because of this similarity, I can transfer my inner understanding outside and my outer understanding inside. Very often this relationship proves important for us because there are certain things that are better understood objectively—that is, in the outer world—and other things that are better understood subjectively—that is, in our own private experience. There are also some things that are better understood through sharing them with others and thus learning about patterns common to different human beings.

The fact that understanding can never be communicated is one reason why men and women can never wholly understand one another, since there are certain kinds of experiences (not merely connected with reproduction) that women can have and that men cannot have, and vice versa. For certain things, a woman's understanding remains womanly and a man's understanding remains manly; there is no way of communicating the woman's experience to a man or the man's experience to a woman. This impossibility of communication illustrates an aspect of understanding that must be accepted and lived with: there is no getting outside of experience. This fact, however, does not mean that experience rightly assimilated and rightly used is the only way that one can gain understanding. Therefore, the question before us is how we can use our experience so that it will increase understanding.

UNDERSTANDING AND SYSTEMATICS

I do not think that I need to discuss the necessity for increasing our understanding. It is understanding that matters when it comes to living, and no amount of knowledge can take the place of understanding. What is not so easily grasped, and perhaps not so easily accepted, is that to understand anything better means to understand everything better. I do not mean here that understanding can be integrally transferred from one situation to another because, in practice, understanding is never completely separable from knowledge. But if we can bring understanding into situations where knowledge plays a relatively small part, then this understanding is of universal value and can be applied to every kind of circumstance.

Herein lies the great value of systematics. It is a means to develop understanding independently, as far as is possible, of any particular kind of knowledge. In studying systematics, you need not have preparation in any subject whatsoever. There is no need to know mathematics, philosophy, science, history, or any other field that might provide illustrative materials. We can work together for understanding, using the most everyday examples that arise in everyone's lives. We can also draw on fields that you know nothing about, yet you may be quite astonished to find yourself understanding much about the subject.

There are different powers that belong to different parts of human nature. Understanding is a deeper power, more intimately connected with our own nature than knowledge, which is much more external to us. This difference is one reason why what we understand is truly our own, whereas what we know may be taken from us. Also, it must be said that understanding is inseparable from some kind of faith. If the world can be understood, it must make sense, and the search for understanding implies—even if unconsciously—faith that there is meaning and significance in the vast array of unknowables that we meet at every turn. To make this faith more explicit and coherent is one task we set ourselves in the discipline of systematics.

SYSTEMS AND STRUCTURES IN SYSTEMATICS

If systematics is the study of underlying patterns common to all kinds of situations, we must first clearly define **system**, which is the term that systematics uses for these underlying patterns. To understand the nature of a system, we can first consider how the word is used in ordinary language. Typically, "system" is defined as an organized whole—a structure made of parts that are connected with one another in such a way that they all enter into the pattern of the whole. For example, we speak of the solar system, referring to a structure consisting of planets and their satellites, meteorites, dust and radiation, all revolving about the sun as center and held together by fields of forces. We call the planets and sun a solar system because our experience shows us that it is an ordered, recognizable pattern. We have very good reason to believe that if a similar group of celestial objects exist elsewhere, it will have a pattern of smaller, lighter objects moving about the larger, heavier ones.

Another use of the word "system" occurs when we speak of the workings of the human body. The nervous system, for example, is a precisely organized structure of different parts connected and related to each other. As research on animal nervous systems becomes more refined, we can transfer much of the understanding to the human nervous system, in both healthy and pathological states. Such connections are possible because we recognize a basic structure in both animals and people—a similar organization of parts that leads us to call this structure a nervous system.

In systematics, a system is similar to these everyday examples in that all its elements are part of an interconnected whole. The system is different, however, in that these elements are generally not obviously discernable as are the movement of the planets or the pattern of synapses and nerve impulses. Rather, the elements of a system in systematics *underlie* the directly given appearance of the thing and are grounded in *the qualitative significance of number.* For example, the monad is a system that uses the qualities of oneness to identify underlying qualities of the thing, while the systems of dyad and triad have the same aim but use the qualities of twoness and threeness. Further, each particular thing—be it an object, a living being, a process, or event—can be explored through several different underlying systems.

We can define a **system**, therefore, as *a set of independent but mutually relevant terms*, in which **term** refers to *those elements of the system that express a specific character,* such as universality, complementarity, dynamism, activity, potential, and so forth. The **order** of the system is given by its number of terms. For example, the **monad** is of the first order and has only one term, which is called **totality. Dyads, triads, tetrads,** and **pentads** are the second-, third-, fourth-, and fifth-order systems. Their respective terms are **natures, impulses, sources,** and **limits.**

One last word that needs definition is **structure**, which in this book is used generically to identify *the particular thing that is to be explored through systematics*. A structure can be an object, a living being, a process, an event, a situation, a human group, an historical era, and so forth. From the vantage point of systematics, *every organized totality is a structure composed of systems*. Structures have a pattern corresponding to one or more of the basic systems.

Typically, no one system alone can elucidate the complexity of real-world structures, and we must draw on several. In the following chapters, we will find that a system of one order may be more useful than another, depending on the nature of the structure or the particular aspect of the structure that we seek to understand. We discover that, for many purposes, a structure can be thoroughly understood by the systems of the first four orders—monad, dyad, triad, and tetrad. If, however, we need to understand why the structure exists and what it is intended for, we must take higher systems into account.[1]

IDENTIFYING SYSTEMS

Before we discuss the first five systems in detail, I want to provide some introductory examples of how one identifies a system. We can begin with the two-term system of the dyad, using a situation that we can recognize from our own experience—dealing with children. If we as adults are to train children properly, our behavior must take care of two contrasting needs: On the one hand, order and discipline; on the other hand, respect for the spontaneity and freedom of the child.

These two contrasting but complementary needs mark out a dyad. On one hand, there can be no real spontaneity without discipline. Freedom without order is not true freedom but, instead, anarchy or completely meaningless behavior. On the other hand, discipline without spontaneity is not order but, instead, a fixed situation in which there is no possibility of anything happening at all. Order and freedom are independent of each other and the action required to produce each is different and separate. To preserve discipline is one thing; to produce spontaneity is something else. The use of the dyad helps us to realize that successful child rearing requires both order and freedom.[2]

As another example, we can consider the human being according to several different systems. A person can be viewed as a physical object, as a living organism, as a complex of activities, as a unique individual, and as much more that we can never know completely. The *monad* of a human being is a reasonably well-defined totality that we can all recognize. In terms of the *dyad*, this totality divides naturally into a visible and invisible part that can be called "body and mind." This dyad cannot be eliminated from any study of the human being, but it does not tell the whole story of human nature.

Quite different from the dyad of body and mind is the division that appears when we draw on the *triad*, which points to three broad types of human experience: bodily, emotional, and intellectual. Gurdjieff argued that these three kinds of activities are located in three "centers": the spinal brain, the feeling brain, and the head brain. From these centers, impulses arise both inwardly and outwardly and produce the complex pattern of human experience and behavior.[3] The greater part of human actions comes from the interplay of these three centers, each of which is necessary to the other two. For example, we can recognize that we can never think unless there is some feeling. If we are totally without interest in a subject, we are unable to think about it. At the same time, our thoughts and feelings are continuously influenced by our physical body—our state of health, our degree of fatigue, and so forth. In terms of a system, the three centers are mutually relevant terms.

At the same time, however, the three centers are independent, since all three sets of functions are crucial to human life. Unfortunately, various human habits often disturb this independence, and one center interferes with another. If we seriously study our own psychological nature, we come to see that the greater the degree of independence among the three centers, the richer and stronger one's life.

When the need arises, this independence allows the centers to coordinate their separate work in an integrated way. The centers of some people are particularly well harmonized, and thinking center can always count upon an agreement and strength from the feelings and the body. Emotional life is in agreement with mental processes and the body provides physical support for the mutually supportive actions of thinking and feeling.

We can also look at human nature from the viewpoint of the four-term system, or *tetrad*. One example is psychologist Carl Jung's typology of human psychic life. He speaks of four terms—intuition, sensation, feeling and thought—that should be completely independent of one another. He argues that these four terms summarize the variety of psychological types and account for the diversity of people's psychic life.

INTEGRATING MULTIPLE PERSPECTIVES

I have chosen the above examples to illustrate a key feature of systematics—that a particular structure like the human being can be viewed in terms of several different systems. From the vantage point of the triad, the person can be described in terms of thinking, feeling and body, whereas from the vantage point of the tetrad, the same person can be considered in terms of intuition, sensation, feeling, and thought. One might wonder which typology is more accurate: Is the person better understood as a threefold or fourfold being? The answer to this question is not an either/or. Rather, systematics illustrates that a particular situation can be considered in terms of several different systems, depending on one's viewpoint and understanding. As Gurdjieff said, people are a three-brained or four-centered being, depending on how one wishes to look. The point is that different systems bring out different aspects of the thing under consideration.

In studying structures from the vantage point of different systems, one must be continuously on guard against the tendency to take one presentation as true and to reject the others as false. In our usual ways of thinking, it may seem that the human being cannot be, at the same time, a twofold, threefold and fourfold being. In fact, this multiplicity is exactly what the person is because of the diversity of human forms and functions, which can be summarized by a relatively small number of systems.

The difficulty is that we must learn to keep the various systems distinct from one another and not force what belongs to one system into another. For example, one should not manipulate Jung's fourfold typology to make it conform to Gurdjieff's threefold system, nor is it right to eliminate the traditional philosophical distinction between body and mind, arguing that it has been superseded by more modern phrasings of human nature. Any structure involves several different systems. Understanding comes when we are able to see how in different ways these systems are all present together.

If this effort to see had to be done separately, in length, for every new structure, systematics would be of limited practical use, since the structure might have changed before we could examine it completely. The great value of systematics is that once we have applied it to a certain number of situations, we begin to recognize how the different systems enter into many other situations and structures.

In fact, I would argue that systematics can help describe *all* situations, provided they are grounded in human experience and not artificially constructed. To be real, understanding must be based on human experience because experience is the one thing that is totally reliable in that it always carries with it various underlying patterns that are unchanging. Things and situations that people invent artificially can never have the completeness of structure that a natural system always has. I do not mean to say here that human inventions are entirely outside the understanding of systematics. On the contrary, one of the most remarkable things is to see that all inventions that really work are built, whether knowingly or unknowingly, according to the underlying patterns identified by systematics.

With this introductory material in mind, we can now discuss the first five systems of systematics in detail. We begin with the *monad*, which is associated with wholeness, universality, and diversity in unity. The monad allows us to delineate the ground of the structure that we wish to study. What does the structure include and what does it not include?

NOTES

1. As Bennett has already explained, the present book discusses systems up to the pentad. For discussion of higher-order systems up to twelve, see John G. Bennett, *The Dramatic Universe, vol. 3: Man and His Nature* (London: Hodder & Stoughton, 1966), pp. 44- 75. Also, see appendix A, "General Systematics," which discusses systems up to eight.

2. The dyad can only identify polarity and complementarity. To understand *reconciliation* between order and spontaneity, one must turn to the triad (see chapter 3).

3. See, for example, Peter D. Ouspensky, *In Search of the Miraculous: Fragments of an Unknown Teaching* (New York: Harcourt Brace & World, 1949).

CHAPTER 1

The Monad

THE MONAD identifies *content*—the special qualities of a thing or situation that make it what it is rather than something else. The monad helps us to recognize *the substantial things of life*—particular entities, experiences, events and processes that make our human world essentially what it is. As I explained in the introduction, **structure** is the generic term for any of these substantial things of life. A structure is any thing, experience, event, process, relationship, and so forth that we can encounter in our lives as human beings. The monad provides a way to describe the content of the particular structure in which we are interested.

IDENTIFYING STRUCTURES

In using systematics, one must understand clearly the nature of structures suitable for study. At first glance, it might seem that anything that can be replaced by the word "something" is a legitimate structure, but this principle is not always true. In Western languages, "somethings" are generally indicated by nouns, but some nouns make unsuitable structures for systematic study because they are too abstract and speak only to one aspect of a thing. For example, "goodness" is not useful as a structure because it is too general and varies for different individuals, cultures, and historical periods.

On the other hand, a structure must not necessarily be some tangible object with clearly defined boundaries. For example, complex entities such as "school," "learning," "government," and "community" can be studied as structures and their content identified by the monad. The key is that the thing must be sufficiently complete in terms of human experience so that we can have contact with it. In this regard, "goodness" is too general and culturally varied to be a structure, while "the battle of Waterloo" is suitable because it is a sufficiently complete event. The same is true for "home" or for a general activity like "music," both of which are shorthand descriptions for particular spheres of human experience.

The next point to realize is that before we begin to understand a "something," we must first *recognize what this "something" is*. We must bring into focus the particular structure we are trying to understand. This requirement is straightforward with material objects; for example, I can point to a painting and say, "We are going to try to understand that picture." Provided you are in a position to see the painting, you are able to recognize the particular "something" we are trying to understand.

There are, however, many things in the world that cannot be illustrated directly—for example, a tree species that lives far away from where we are, or a structure, such as "school", that involves less tangible qualities like learning and responsibility. In these cases, there is no "something" that one can easily point to and say, "This is what we will try to understand."

For the majority of these cases, however, we at least have words, and one important use of language is to enable people to identify structures that would otherwise be indescribable. For example, if I say, "Let's talk about trees," the word "trees" conveniently substitutes for my taking you out into a wood and saying, "This, this, and this are the kind of thing we are going to discuss."

The difficulty with this shorthand step of language, however, is that we may not all recognize the same object when we use a particular word. There is little difficulty for readily clarified words like "tree," but we often face more difficult problems when we leave the material realm and consider less tangible structures. There are not nearly enough words to designate each "something" separately, and even if there were, we would be overwhelmed in keeping the many things and associated words in order. Instead, we learn to use words in an approximate way. If we look at a typical dictionary entry, for example, we notice that there are several meanings for a word. The same word may have a different meaning for different people, and many misunderstandings arise from this inability of words to evoke exactly the same sense for each person who uses them.

For everyday purposes, the imprecision of language is not a problem because the context of communication—voice inflection, bodily gestures, speakers who know each other, and so forth—contributes to clarification and understanding. This solution, however, is inadequate for systematics in that we cannot designate the particular "somethings" to be studied just by using a word and hoping that we shall all recognize the same thing. This dilemma not only involves efforts of collective understanding. Even when we work alone, we must be certain that we understand precisely the words we use.

Whenever we set about to understand a particular "something," whether singly or as a group, we must first make absolutely certain that we are clear about what the "something" is. If people are working as a group, they must verify the "something" by, first, sharing their individual understandings; and, second, arriving at a consensus as to what the "something" is as the group will understand it. This stage of consideration requires a discipline that is often neglected, yet this stage is particularly important in group study. Too often, groups do not carefully establish an agreement as to the focus of their subject, and the result is misunderstanding and wasted time.

RECOGNIZING MONADS

Understanding a structure requires openness and an effort to see clearly and completely. At the very start of our study, we must make absolutely certain that we know the content of the structure in which we are interested. We must not rely on a hope that the content will become clear as we proceed. This effort to look and to see is the first step for understanding a structure. In the language of systematics, this step is *recognizing the monad*—that is, identifying the particular universe that we wish to study. Every monad is a universe in itself and has a diverse content drawn together by a central quality that makes the monad a recognizable totality.

In part, this gathering quality of the monad allows human beings to have language. From an early age, we give names to things and experiences. Each time we face a new problem of understanding we must do as we did as children: search to distinguish patterns in a diversity that is seemingly chaotic. The monad provides a self-conscious means for discovering and identifying patterns that in our everyday life we might never see or that we typically see only in partial and haphazard fashion.

At first, it may seem that the monad is nothing more than *one*. For example, I might say that a piece of chalk is a monad because it is just one object and nothing else. But the monad is something more than the oneness of an object. Rather, the monad is what we immediately contact in our experience before we have understood anything about a thing or "made anything of it." Imagine yourself a new-born baby who opens her eyes and sees the world. She does not make anything of what she sees and does not distinguish one part of the world from another. There are no objects, ideas or other particulars; instead, there is only immediate experience. More than likely, the baby sees a set of blurred and confusing images that are not differentiated. This set of images, however, is not uniform. There are portions of light and darkness, for example, and the baby closes her eyes if the light is too strong or turns her head toward light that is more pleasant. The baby's experience is *diversified* but not *differentiated*.

To differentiate means to distinguish one thing from another and to be aware of differences. Awareness of diversity is more rudimentary than awareness of differentiation. Recognition of difference is not yet the ability to distinguish one thing from another. Suppose, for example, that a man born blind suddenly gains sight. What does he see in the first moment of vision? More than likely, he does not see everything as the same, but he cannot distinguish one thing from another. The Gospels tell the story of a man who suddenly regains his sight. He describes people "as trees walking." Though this man can differentiate between people and trees by touch, he cannot by sight. Quickly, however, he realizes the nature of visual distinctions and is able to distinguish one thing from another.

The blind man's exceptional case is not the only way to illustrate the difference between diversity and differentiation. One ordinary example involves those occasional moments between sleep and waking when we are conscious but not aware of anything in particular. Quickly, however, our more conscious self begins to recognize things, and the typical everyday order of our waking world returns.

What importance does this primitive experience of pre-awareness have for our lives, since most of the time we live with distinctions that matter? In fact, our powers of understanding are greatly limited exactly because of the habit, formed at a very early age, of our experiencing distinctions without first seeing diversity. There is the adage, "See everything as if you had never seen it before"—surely a wonderful piece of advice, but how can it be heeded? In fact, one *can* with practice. There are certain exercises, both psychological and spiritual, that can help one to see things as if he or she had never seen them before. Anyone who has seriously practiced exercises of this sort is astonished at the transformation and enrichment of experience that these exercises bring. One realizes that the leap from immediate experience to differentiation causes him or her to miss something extremely vital.

We can, without any special exercises, have immediate contact with the world in front of us simply by approaching it so that we do not start with distinctions. This technique is the heart of identifying a structure as a monad. A simple way of describing this technique is to say that when we wish to understand a situation, the first thing we do is simply to collect all the material we can about the situation without first attempting to make anything of it. Let it come just as it comes. In this sense, we are identifying the content of the structure and describing it as a monad. We set our structure before us and allow a description to come forth.

One way to carry out this procedure practically is to make a list of qualities that characterize the particular structure, making sure that we do not order these qualities or draw links among them. In other words, conceptualization is not appropriate for the monad; one should not classify elements or give them intellectual shape. An effort of openness, however, does not mean a lack of direction. The guiding principle for the monad is *whether a particular element is relevant or not.* To know what is relevant, one must have some sense of what belongs to the structure or have that sense of belonging conveyed vicariously.

Working for openness toward the thing we wish to understand is certainly not a technique unique to systematics. There are many people who have understood the importance of this step for all kinds of mental training. Everyone, when confronted with a new problem, must first collect the contents of the situation before carrying out an action, whether that action is assigning places to things, analyzing, synthesizing, or drawing conclusions. To identify the content of a structure is crucial, and I have found that if this step is neglected in systematics, consideration of any higher-order systems becomes artificial.

A SCHOOL AS AN EXAMPLE OF THE MONAD

To illustrate how the monad can be used to identify the contents of a structure, let me provide an example. Suppose that the situation I want to study is a school. In this case, the school becomes our structure to be interpreted through the monad. What we must do is to hold an image of "school" before us and to record the various elements that appear. We immediately visualize pupils, teachers, classrooms, janitors, equipment, curricula, lessons, examinations, organization of classes, relationships among pupils, teachers and parents, and so forth. These elements and many more quickly come to mind.

This visualization procedure may seem unnecessary but, in fact, it is crucial because it *allows us to bring forth the content of "school" without translating this content into any kind of mental construction.* This unshaped content of school is its monad. In turn, this monad is a *universe*, which always involves diversity in unity. The various elements of this monad are connected and unified by the fact that they all refer to the situation called "school." On the other hand, these elements are not only different from one another but also of diverse natures: some are material objects (buildings and furniture), some are people (pupils, teachers and parents), some are activities (teaching and learning), some are directives (the curricula and examinations), and so on. Diversity is not only a difference among things of the same kind but also a difference among things of different kinds. These things involve not only tangible entities like equipment and participants but also less readily identifiable entities like learning, discipline, and so forth. The totality of all these things is the content of the monad.

THE SHIFT FROM MONAD TO DYAD

The monad identifies content, but it cannot identify patterns within that content. Only higher-order systems can help us with that aim. To illustrate the transition from content to pattern, I want to examine the monad of "home" and then illustrate the shift to the home's second-order system—its *dyad.* Chapter 2 explores the dyad in detail and, to prepare for that discussion, it is useful to consider the transition from monad to dyad, which in turn involves the transition from identity to difference.

To identify the monad for home, we can follow the same procedure that we used for the monad of school. We make a list of descriptive words or phrases that identify the contents of a home's universe—for example, house, food, kitchen, living room, furniture, sleeping, parents, children, shelter, privacy, hospitality, and so forth. In reading through this list, we might sooner or later realize that the elements can be broken into various subgroups. For example, one notes that some elements of the home monad are tangible while other elements are intangible. There are material contents such as building, rooms, equipment, family members and their activities; and less observable qualities such as shelter, harmony, privacy, and companionship. Though this division of the home's tangible and intangible elements arises from the monad, it does not really provide a meaningful pattern for the home. Placing children in the same category as kitchen and furniture, or separating harmony and protection from the people having these experiences, does little to help us understand the home. In short, *identification of content does not guarantee understanding of pattern.*

Yet the higher-order systems involve pattern, and we must find a way to discover pattern in regard to the order of the particular system with which we look at the structure. For the dyad, one typically finds that an important pattern is the tension between the structure itself and the structure's relationship with the larger world. In regard to home, for example, one set of descriptions seems to emphasize the home's self-contained qualities—food, kitchen, furniture, parents, children, privacy, shelter, contentment, and the like. In contrast, another group of descriptions highlight the home's outside-oriented qualities—hospitality, a base for coming and going, a place to get ready to meet the world, and so forth.

These contrasting sets of elements point to an inescapable quality of home: its dual function as an inward and outward place. When we mention family, shelter, and comfort, we speak of the home's powers to gather and hold. But when we speak of hospitality or departure and return, we speak of the home's powers to relate its dwellers to the larger world. Because we have a home, we have a *place in the world.* Homelessness is not mere lack of protection; it is *to be without a place.* In this sense, the home is both a place that concentrates our lives and also a place that we occupy in relation to the larger world.

"Place," therefore, has two quite different meanings. In one sense, it is the "place" into which we enter, but in another sense, it is our "place" from which we move into the world. These two meanings permeate the content of a home and point toward its significance as a dyad: Some elements relate to the home as a place of concentration and personal identity, while other elements relate to the home as a place of expansion that joins its dwellers with the larger society.

To speak of the home in these two ways, therefore, is to do more than to simply describe content. The polar nature of home points toward an essential pattern and speaks to the *meaning* of the home and *what it stands for.* The aspects of inside and outside belong always and already together when we consider the home. On one hand, the home is a supportive, receptive place where we can feel open and able to welcome in the outside world. On the other hand, the home is a closed, protective place that keeps out the larger world. In fact, when we consider any structure, we realize that its contextual meaning is different from its meaning as a thing in its own right. The question always arises: *When we consider a structure, how are we to understand it as both a thing in itself and also as part of a larger world?*

If these inside and outside dimensions are not considered together, we do not have a complete picture of the structure. Looked at from one dimension only, the home is no longer really itself. A home that continually shuts out the world destroys the fullness of human life, just as the home offering no shelter from the larger world destroys family identity, as the members lose a life of their own. The meaning of home always includes a conflict of demands between centripetal and centrifugal tendencies. From one angle, these tendencies might be described as "feminine" and "masculine," though these descriptions are figurative and point to qualities of the home, not gender-specific roles and obligations. The feminine tendency involves closure and the needs of protection and nurture, while the masculine tendency sees the home as a door to the world of enterprise, challenge, and wider horizons.

In nearly all the elements of a home, one can find this division of interests. For example, a cooking utensil, from one point of view, is a symbol of nourishment and family meals. From another angle, however, that utensil helps the house welcome the world and is, therefore, a symbol of hospitality. At times, a chair provides a private place where a mother nurses her baby, yet at other times that chair welcomes a guest to sit and feel at home. Yet again, a family shares its roof with others, but that same roof offers protection from the elements.

In giving attention to the fundamental division of a structure, one moves from the monad (the home as an undivided whole) to the dyad (the home as a twofold). At first glance, this division may seem unique to the home and not readily transferable to other structures that do not have the central and peripheral qualities of home. In the next chapter, however, we shall see that many dyads involve this complementarity between what the structure is in itself and what it is in relation to its world. In one sense, this complementarity points to the two natures of any structure: *what it is* and *what it does*. These two natures arise from the monad but do not supersede it. Rather, these two natures move us to the dyad, which is the focus of chapter 2.[1]

NOTE

1. In *The Dramatic Universe*, volume 3, p. 19, Bennett says the following about the relationship between monad and dyad: "What [a structure] is, that is the content, is its own affair; but what it does concerns everything around it. There is no end to the repercussions of the smallest act—even the splitting of an atom. Every monad—being the form of a structure—bears within it the two-fold significance of its source. It is infinite in its external connectedness, and it is also infinite in its internal diversity. The two infinities are not the same. They even contradict one another. The inner significance comes from separation from the rest of the world and the outer from contact with it. This can easily be seen in any actual monad: for example, a home. We even go so far as to say that there are always two homes: the mother's which draws in and the father's which reaches out. And yet both homes are the same home—the same monad. Such considerations point to the suggestion that understanding cannot stop at grasping the monad: it must go on to face the dyad. By definition, a dyad is a two-term system, such that each term is distinct from and yet requires, and even presupposes, the other."

CHAPTER 2

The Dyad

A PERSON can experience things in two ways: from within and from without. To draw a circle is to mark the boundary of a thing symbolically and to describe that thing as it is unto itself. The spatial content within the boundary is limited, while the spatial content outside is infinite and symbolizes the thing's participation with the world outside itself.

In understanding anything, one finds that there is always one perspective that sees the thing in itself and another perspective that sees the thing in terms of everything that it is not. One realizes that a complete picture of the thing requires that it be understood in relation both to what it is and to what it is not. All that is outside the thing has meaning and sheds light on what the thing itself is. This division is not logical or arbitrary but absolutely necessary for understanding the thing.

If, for instance, we return to the structure of the home discussed in chapter 1, we realize that all that is outside the home is necessary for understanding the home itself. The home is not an isolated, self-contained universe but a living part of human society. Its materials are all part of the earth's surface; its members come from the larger world and return to it; the home arises from a past and moves into a future. In this sense, the aspect of the home that looks outward is just as necessary as the aspect that looks inward. This outwardly-directed quality is as much a part of the home as its inwardly-directed quality.

These two aspects of the home do not exist because of our shifting perspectives as observers; nor do we arbitrarily impose these two aspects through a predefined mental picture of what the home is. Rather, these two aspects are *inherently in* the home. Instead of mere appearance, these inner and outer aspects are an integral part of all things. From the vantage point of systematics, these two aspects refer to the two terms of the dyad, each of which we describe by the word **nature**.[1]

In this way, we say that a home has two natures, whether they be described as inner and outer or private and public. These two natures do not divide the home into two parts. One cannot say, "This part belongs to the private nature, while this part belongs to the public nature." This statement cannot be true, because if it were, we would have two things rather than one. The home would simply divide into two parts, each of which would then become a new thing in itself.

The two natures of anything are not two parts of it. Instead, these natures are *pervasive* in the sense that they intimately permeate every portion of the thing, even the parts to which they seem opposite.[2] Again, we can consider the home, in regard to which there is nothing whatever that is not concerned with its nature as a private place. At one moment, the door and chairs welcome guests but, at other moments, secure the home from the public world and contribute to a place of retreat. Yet there is also nothing in the private nature of the home that is not related to its other nature as a place that belongs to the world. Even the most protective parts of the home contribute to its ability to play a public role, particularly as these parts help renew family members and give them restored energies to face again the outside world.

If the home were to disregard its public obligation, it would fall into isolation, and the character of its every part would change accordingly. Similarly, the home that turns itself entirely toward its public role is also radically changed. We have all experienced homes that exaggerate one or the other of their two natures—the home that seems to exist only for visitors, or the home that turns its back to the world and feels inhospitable. Because of its integral polar tension relating to its character as a dyad, the home faces a fundamental conflict of demands. If its two natures lose their distinctiveness or fall into open conflict, the home itself collapses.

In symbolic terms, the private and public natures of the home are associated with feminine and masculine qualities. Because of this association, the family members who represent these qualities often understand the objects, needs, and purposes of the home in entirely different ways.

Unless one nature of the home completely subordinates itself to the other, a single individual can never have a complete awareness of the home's two natures. For the home or any other structure, we must accept the impossibility of full understanding. All structures present themselves in terms of two natures that we cannot determine fully without destroying the structures themselves. Complete determination of the two natures would break the wholeness of the structure into two artificial parts different from the original monad.

THE SHIP AS A DYAD

To satisfy ourselves that every structure has two natures, we need to work through many illustrations from our own experience. One example I want to consider is a ship, which as a monad includes passengers, freight, a route from port to port, a relationship with the sea, and so forth. Note that these aspects of the monad describe the ship's *relation to the world*—the outwardly-directed nature of the ship. In addition, however, there is the ship *as a ship*—its size, its design, its materials of construction, its crew, and so forth. In this sense, the ship has its own existence and identity—its own special nature.

The way that a ship finds itself and takes on a unique identity is an extraordinary experience. Though I have never seen a sailing ship go through trials, I once was on the test run of a steamship that was going out into the Indian trade. To see a ship launched and to participate in its trial is to experience a coming to presence of a thing that is not only metal and machinery, but a being that will live in its own way for many years. One feels that not only the engineers have passed the trials but also the ship itself. This quality of *in-itselfness* marks the ship's inwardly-directed nature. Though these trials took place over thirty years ago, I know the ship still sails and I still feel a link with it.[3]

But the ship's intimate nature is only one part of the dyad. The ship could not find its unique self if there were no sea and storms to face or no freight and travellers to carry. The Indian ship carried cargo and third-class passengers to ports in Malaya, India, and Burma. Even in the trials, one already felt the ship's future life. This outwardly-directed nature of the ship is also quite real and relates to the ship as it does its work and exists in the world of sailing, transportation, and commerce. In most general terms, the ship exists as a part of human life on the earth, and in this way expresses its public nature.

The crucial point to understand is that in considering the ship as a dyad, we do not divide it into two parts and say that what is below the Plimsoll line is the inside, or the intimate ship, and what is above is the outside, or world-directed ship. Nor do we say that inside the hull is the inner ship and outside the hull is the outer ship. The dyad is not like that. Instead, *the whole ship* has two different natures: the ship's own intimate nature and its outward or public nature. Nor is it correct to say that one nature is the face the ship presents to the world and the other is the ship as it itself is and what it means for the crew. The dyad cannot be articulated so exactly because its two natures permeate each other everywhere.

A SCHOOL AND ORCHESTRA AS DYADS

What is the practical value of the dyad? In our lives, we face the dyad continually, but because we are not aware of it, we typically turn our attention one way or the other and thereby fracture the indivisibility of the whole. In practice, it is extremely difficult to hold the two natures of the dyad together, and systematics is valuable in this regard because it provides a way to see the two natures and perhaps through this seeing to better work with and hold them together.

We can perhaps better realize this practical value of the dyad by considering a structure that is a social institution—a school. If we consider the two natures of a school, we realize that its intimate, inner aspect involves the development of children as individuals who must be able to read, to write, and, overall, to become good citizens.

At the same time, the school has a relationship to the world beyond, and school officials must uphold the formal policies authorizing the school, deal with parents and other outside people involved with the school in various ways, stay in touch with the wider educational system of which the school is a part, and so forth. These external responsibilities of the school often seem painfully removed from the school's intimate action upon the individual pupil, yet these obligations are integral and must be considered if one is to understand the school fully.

If, on the one hand, we give attention only to the pupils' learning, we shut ourselves in an ivory tower built from our own incomplete beliefs as to what the perfect education is. If, on the other hand, we think of the school only in terms of administrative and economic requirements, we pay only lip service to the intimate needs of individual pupils and become lost in the politics of education versus society. The person truly desirous of educational quality must face the conflicting demands inescapably present in the school's dyadic character.

We can also see these conflicting requirements if we consider an orchestra as a dyad. An orchestra generally faces two competing demands. On the one hand, it is a set of individual musicians who through talent, dedication, and practice come to express a group style and character. In this sense, the experience of an orchestra's finding its identity is in many ways similar to the earlier example of the ship. On the other hand, and also like the ship and school, the orchestra must face a larger world that includes financial funding and public reaction. Like any other cultural institution, the orchestra must support itself economically.

We see here the tension between the two natures of the orchestra—the need to be genuine artistically versus the need to survive financially. Anyone who has played in an orchestra knows firsthand the tension between these two needs and the artistic sacrifices required for even the most fortunate orchestra to accommodate itself to economic and social demands. The orchestra must give a certain number of concerts prepared for by a limited number of rehearsals. The conductor must reconcile his or her personal tastes and wishes with the tastes and wishes of the public.

There is no easy balance between the two natures of the orchestra. The director must maintain the orchestra's artistic integrity yet, at the same time, protect the orchestra's economic soundness. There can be no compromise for either of these two natures. And here lies the crucial significance of the dyad: *there can never be true satisfaction if one or the other of the dyad's natures are given short shrift.* If the orchestra sacrifices artistic integrity or if it does not pay for itself, then it is not its real self and will sooner or later collapse.

THE IMPOSSIBILITY OF RESOLUTION

The dyad demonstrates that the orchestra's two demands—musical quality and economic viability—are inexorable yet completely different from each other. One cannot say, "Well, it does not so much matter that we have not rehearsed this piece thoroughly because, after all, we have a good box office." At the same time, he or she cannot say, "Well, we cannot pay our debts, but, by Jove, we have kept our artistic integrity." The two terms of the dyad are such that each has its own absolute right to exist, and a lack of one cannot be offset by a greater presence of the other.

In the dyad, it is impossible to balance the two terms because a balance assumes some middle point where the demands of the two sides are blended and satisfied. The polar qualities of the dyad *by their very nature* are irreconcilable and cannot be brought into balance. In ordinary life, we ignore this impossibility and speak instead of some middle way that can lead to resolution. But in speaking of a middle way, we confuse the dyad with the triad, which is the real vehicle for reconciliation.[4]

In other words, a middle way is different from the dyad's contrasting natures, neither of which will accept anything less than perfection. If we return to the dyad of the school, for example, we recognize that, on one hand, there can be no educational compromise in regard to the individual child. Each pupil has the absolute right to the best possible education. On the other hand, society also has an absolute right to demand good citizens. We are all social beings, and education must contribute to the integration of individuals into society.

In practice, a school seeks a compromise by developing both needs as well as it can. The dilemma is that the demands are entirely opposite and will not admit compromise. Too often, a school emphasizes one nature at the expense of the other, and the school begins to falter. For instance, in modern Western schools, the concern with crime, drugs and community integration often overwhelms the process of education and attention to pupils' individual well-being.

Another example to illustrate the irreconcilable quality of the dyad is a city. Like a ship, a city must find itself, and each city requires a certain amount of time to reach its maturity. If one visits an ancient city and a youthful city, he or she notes that their sense of place is very much different. Older cities like London, Rome or Istanbul express a much different atmosphere and character than newer cities like Los Angeles, Brasilia, or Chandigarh.

Whether the city is young or old, its unique ambience relates to the dyad's intimate nature—that quality that identifies the city's individuality. This individuality also relates to the power that a city has to hold itself together and to cause people to say, "I am a citizen of that city." At the same time, however, there is the city's relationship with its larger geographical context—its commerce and traffic, its political and economic links, and so forth. These wider bonds mark the outside, contextual nature of the city and often conflict with the needs that the city has as a place in itself. But whether in regard to unique identity (inner nature) or relationship with surroundings (outer nature), the place about which we speak is the *same* city, just as we speak about the same ship, school, or orchestra.

It is very difficult to understand the dilemma of the dyad's expressing both sameness and difference simultaneously. Perhaps this irreconcilable quality can be clarified further by discussing another example in the realm of the physical world—England's Salisbury Cathedral, one of the great works of Gothic architecture. As a dyad, this building is, on the one hand, a place of worship and, on the other hand, a work of art. The cathedral speaks for God yet is a human achievement.

At first glance, these two natures may seem perfect, each in its own way. If so, the Cathedral would be a dyad that did not involve conflict. In fact, the two natures of the Cathedral cannot resolve themselves, and this irreconcilable difference points to an important point in regard to the relationship between art and religion: each demands a perfection that is quite contrary to the other. One can observe this conflict by a study of world history: periods of great artistic achievement are generally followed by a decline in religious experience. This pattern, in both Western and non-Western cultures, indicates that art and religion eventually face irreconcilable differences.

I remember that many people were shocked when the Shivapuri Baba, a remarkable Indian holy man, spoke of the conflict between art and religion: "Art will only take you to beauty, not to God."[5] This conflict does not mean that one must choose once and for all between art and religion. The dyad does not work in this way because, as has been already said, if either of its sides are overemphasized, the dyad disappears.

If one accepts the incompatibility between art and religion, yet also realizes that art and religion are both necessary in people's lives, he or she begins to understand the dyad. One no longer says, "I must abandon art because religion is more important," or the opposite. Instead, one faces the fact that art and religion make opposite demands and one must bear this conflict. In fact, Salisbury Cathedral was built in the thirteenth century, a remarkable time in the way that people did bear a number of very sharp conflicts from which arose many powerful experiences that would affect the course of European history.

The dyad is an unavoidable part of reality and contributes to the nature and outcome of every human situation. There is a starkness in the dyad that must be accepted, since its irreconcilable opposites cannot be otherwise. Any compromise between the dyad's inner and outer natures presupposes a relationship between them, but there can be no dyadic relationship because relationship is a property of the triad. The better our understanding of the dyad, the more our strength to bear polarity and to seek reconciliation through the *triad*, which involves relationship and dynamism. We turn to these topics in chapter 3.

NOTES

1. Later, in *The Dramatic Universe*, vol. 3, Bennett called the terms of the dyad *poles* and designated them as *positive* and *negative* (p. 20).

2. On the relationship between parts and whole, see Henri Bortoft's work listed in appendix B, "Further Reading."

3. "A ship's nature as a ship is wonderfully described in Kipling's story, 'The Ship That Found Herself'. Everything that one can say about the ship as having its own existence is portrayed in this story." See Rudyard Kipling, "The Ship That Found Itself," in *The Day's Work* (London: Macmillan & Company, 1964, pp. 78-101 [originally 1898]). As the captain of a newly-built steamer explains to the owner's young daughter who has just christened the ship: "...it takes more than christening to make a ship. In the nature of things... she's just irons and rivets and plates put into the form of a ship. She has to find herself.... She's all here, but the parts of her have not learned to work together yet. They've had no chance....Every inch of her...has to be livened up and made to work with its neighbors—'sweetenin' her, we call it technically" (pp. 79-80). Through personifying the various parts of the ship, Kipling describes in the rest of the story how this "sweetenin'" happens during the steamer's maiden voyage.

4. In *The Dramatic Universe*, vol. 2, p. 93, Bennett gives a striking example of the difference between the dyad and the triad: "One man, importuned by a beggar, sees that he is hungry and, acting from the morally good impulse of generosity, gives him money on which he gets drunk and commits a crime. Another man, in the same situation, recognizes that the beggar is a drunkard, and, seeing that he is really hungry, takes him and gives him food, but not money.... The first situation is dyadic; good and evil are contradictory and yet complementary. The beggar goes inexorably to his fate and the generous man is the blind instrument of his downfall. The second situation is triadic. Compassion takes the

place of generosity and a relationship is established, the dynamism of which may save the beggar from his fate."

5. See John G. Bennett, *Long Pilgrimage: The Life and Teaching of the Shivapuri Baba* (London: Turnstone Press, 1975), pp. 164-165. The full conversation is as follows:

J. G. Bennett: May I put to you some questions sent by a group of friends in England? The first is from an old gentleman, Mr. K. He asks: "What value has art compared to charity? Through beautiful things, such as music and painting, we feel ourselves drawn to God. Is art, then, something valuable for man, or what position does it occupy in our lives?"

Shivapuri Baba: One is never drawn toward God by art and music: we are drawn to beauty only. Yes. One is never drawn toward God.

JGB: Art lifts up the soul. Is that not towards God?

SB: No. It will only give you the excellent beauty of God—not God himself.

JGB: But beauty comes from God also, does it not?

SB: But now you see it from a distance. [He holds out the rose he is holding in his hand.] This is beautiful. Can you have the smell?

JGB: Not from here.

SB: The Beauty of God you can know through art and music. But God you cannot know.

JGB: Would you say then, Babaji, that from the point of view of coming to the knowledge of God, art is no longer valuable for man? [To appreciate the emphatic answer it would be necessary to hear the vibrations of the world "harmful" as he pronounced it. It gave an impression of the danger of being led astray by the love of beauty.]

SB: Not only not valuable—harmful also! Mind will not empty of all its contents. Yes! The more beautiful, the further away from God. Yes! Ugliness and beauty: both must vanish from the mind.

Marjorie von Harten [one of Bennett's students]: But, is there not some art which seems much higher than others, like the Sphinx, the Taj Mahal, or the Elephanta Caves with the great Trimurti. That art seems to appeal to another side of us.

SB: But at that time when you see them, you forget yourself, you forget God! Then that beauty alone prevails in you. The essential things are forgotten. Individuality is forgotten. God is forgotten. That beauty only prevails in you at that time. What is the advantage here? I will tell you. Every trouble of the world is gone. One is very happy here. That is all. The unhappiness of this world is not felt.

JGB: But sometimes it is the opposite. There are some religious subjects—they are very common in Christian Art, and also some music, such as the Gregorian Chants for the Passion, that bring us into a vivid sense of the suffering of the world. There is no forgetting unhappiness here.

SB: No. It is only beauty. There may actually be a relationship and yet it is only beauty that prevails in the mind.

CHAPTER 3

The Triad

THE THIRD-ORDER SYSTEM is the *triad*, which relates to action, relationship, and the life of a situation. If we understand the nature of relationships, we can understand why they break down and what must be done to bring them back to life. We realize that relationships are always interconnected and that a shift in one may lead to shifts in others. Without an understanding of the triad, it is difficult to make any real change in the world.

In chapter 2, the question arose of reconciling the dyad's conflicting demands by finding a middle way. This possibility, I said, belonged to the triad rather than the dyad. Before we consider what a middle way might be, we first must examine the nature of relationship. What is a relationship and how is it linked with threeness and the triad?

THE NATURE OF RELATIONSHIP

Two things lying side by side are not involved in relationship but in *connection*—an arbitrary link based on accidental contact that involves no deeper bonds of mutual involvement or concern. For example, I break my leg and you happen to be my hospital nurse for three days. We interact only minimally, and the bond between us is limited to connection. If my break is severe, however, and you nurse me for many months, we may feel a growing fondness and concern for each other. We form a *relationship*.

A relationship requires *exchange and mutual involvement*. More precisely, we can say that a relationship involves four characteristics—dynamism, reciprocity, difference, and reconciliation. First, a relationship is *dynamic* rather than static. One does not necessarily establish a relationship just by placing two things side by side. There must be an exchange between them, and this exchange indicates a mutual action that is continuously at work, whether physically, emotionally, intellectually, or in some other way. A second characteristic of relationship is *reciprocity*. If you and I are to have a relationship, I must participate and have a share in you, and you must participate and have a share in me. This participation must be mutual in the sense that I cannot involve myself in you if you do not involve yourself in me.

A third characteristic of relationship is *difference*. Two identical things cannot have a relationship because they are exactly the same and have nothing to give in the way of reciprocity. In other words, things must have something different to offer each other if they are to be related. Suppose that two men both have horses. The men cannot make a trade because they both have the same animal. But if one man has a horse and the other a cow, a trade is possible if the exchange is mutually valuable. The ability to offer difference is the very essence of a relationship. In this sense, a man and a woman or a weak person and a strong person offer each other different things and therefore have the potential for relationship. In the same way, giving and taking can lead to relationship, but not giving and giving or taking and taking.

A fourth characteristic of relationship is *reconciliation*. Two different things can have a relationship only if they can resolve their dissimilarity in some way. Crucially, the differences must be reconcilable; if they are not, the two parties can never participate in each other. Even when the two things are in powerful opposition, however, a relationship is still possible, providing the opposites are resolvable. There is the adage that hate is near to love. This saying is so because with hate there is possibility of reconciliation.

THE THREE IMPULSES AND TERMS OF THE TRIAD

Looking at relationship as an exchange, one comes to the significance of three: besides the two parties participating in the exchange, there must also be some medium of exchange. In terms of systematics, this medium is the third component that moves us away from the polarity of the dyad toward the reconciliation of the triad. In *The Dramatic Universe*, I used the word **impulse** to identify the three components of the triad.[1] When

Gurdjieff first introduced the triad in his Law of Threefoldness, the three components were translated from the Russian as *forces*.[2] I prefer *impulse* because the word connotes something going out of itself, whereas *force* connotes something that remains within itself. *Force* better describes the dyad, which can never be resolved, while *impulse* better describes the triad's qualities of dynamism, action, participation, and reconciliation.

To identify the three impulses of the triad, I use the words **affirming**, for the impulse that gives or acts; **receptive**, for the impulse that receives, is acted upon, or resists; and **reconciling**, for the impulse that draws the other two impulses together. For the sake of brevity, we can simplify these labels further and speak of the **first** (affirming), **second** (receptive or denying), and **third** (reconciling) impulses, or, most simply, *1*, *2*, and *3*. At the same time, these impulses can occupy three different positions in the triad. We call these positions **terms** and designate them as follows: the first position is the **initiating term**; the second position, the **characterizing term**; and the third position, the **outcome**. The impulses and terms are summarized in table 3.1.

Table 3.1   The Impulses and Terms of a Triad

The Three Impulses
1—affirming impulse (active or initiating)
2—receptive impulse (passive, receiving, resisting, or denying)
3—reconciling impulse

The Three Terms
( ) ⟶ ( ) ⟶ ( )
initiating term   characterizing term   outcome

With this terminology, we can characterize a triad precisely. Any of the three impulses of a triad can occupy the three terms, thus we can speak of a *1—2—3* triad, but we can also speak of a *1—3—2* triad or a *3—2—1* triad. If we work out all the possible combinations, we find that there are six different triads, as illustrated in table 3.2, which arranges the six triads according to the impulse that occupies the initiating term. Later in this chapter, we will discover that each triad represents a different relationship, meaning, and quality. We will also discover that these triads represent the six universal actions and relationships in the world.

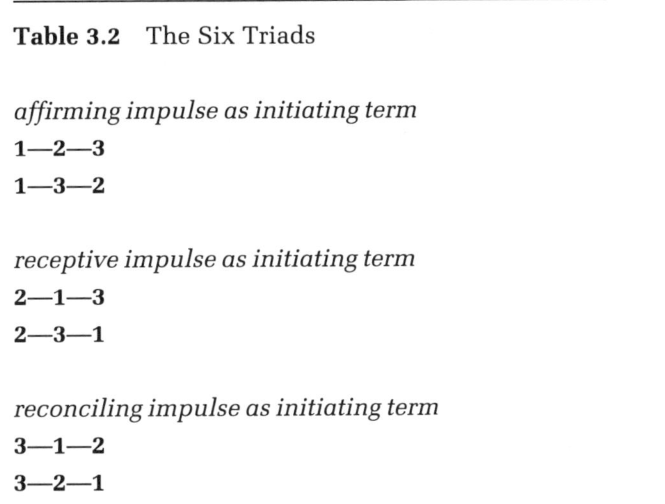

Table 3.2 The Six Triads

affirming impulse as initiating term
1—2—3
1—3—2

receptive impulse as initiating term
2—1—3
2—3—1

reconciling impulse as initiating term
3—1—2
3—2—1

THE TRIAD OF MARRIAGE

Before we discuss the six triads in detail, it is useful to examine some specific relationships as they can be interpreted in terms of the triad. We will make some reference to the six triads above, but their full explication will come later, once we have understood some of the subtleties and complexities one finds in working with relationships and triads.

As one example of the triad, consider marriage. If a man and woman are to be successful as husband and wife, various actions must happen between them. For instance, they set up a home; they have children; they become an inspiration for each other's work. Each of these actions—a home, children, mutual inspiration—provides a reconciling link between the masculine and feminine natures. In this way, there is a movement away from the dyad of man and woman toward the triad of marriage, relationship, and mutual involvement.

This example of marriage also illustrates that triads are not simple or static. Marriage is the triad of husband-wife-home, but it is also the triads of husband-wife-children and husband-wife-work. Marriage involves a host of relationships and is thus a complex network of triads. In addition, the three impulses of a triad can assume different roles. The home, for example, is a connecting link between husband and wife. At the same time, however, the wife provides a home for her husband, who is thus related to the home through his wife. Yet again, the husband provides a home for his wife, who is thus related to the home through her husband. In one sense, the home is the link between husband and wife, yet in another sense, the wife is a link between husband and home, and the husband a link between wife and home. We shall shortly find that the six triads identified above help us to understand these various links more thoroughly.

In fact, we shall discover that these six triads help to explain the extraordinary wealth of relationships in the world. A thing or event can be a part of many different triads and relationships. For example, a family is a triad, but it can also be an impulse in larger societal triads—for instance, those dealing with education or government. That which is a triad at one level can be an impulse in a triad of a higher level. The whole of life is a nested hierarchy of impulses, triads, and relationships all interconnected in a complex network.

OTHER EXAMPLES OF TRIADS

The complexity of human events is seen even in such a simple act as a teacher's writing on the blackboard. Here, the teacher, having a meaning to express, represents the affirming impulse, while the blackboard, receiving and holding the meaning, represents the receptive impulse. The chalk, in bringing about the actual event of writing, represents the reconciling impulse. This triad of writing, however, is part of a larger triad in which the teacher seeks to establish a relationship with his students. His writing on the board is a link between his teaching and the students' learning. At the same time, the words written on the board carry meaning and, therefore, have a certain independent power of their own to relate teacher and students.

We also must understand that an action like writing on the blackboard or making a home can be represented by one of the six triads of table 3.2. Consider, as an example, the action of giving and taking. Here, the donor is the affirming impulse (*1*); the receiver, the receptive impulse (*2*); and the gift, the reconciling impulse (*3*), since it enables the act of giving and taking to be accomplished and the donor and the receiver thereby to be related. Clearly, there is no relationship if the receiver refuses the donor's gift.

The giving-receiving relationship can be summarized by the triad *1—3—2*, which says that the affirming impulse acts on the receptive impulse through the intermediary of the reconciling impulse. This triad not only describes the relationship of giving and taking. It enters into many situations where something is accomplished or changed. For example, the affirming impulse of the wind (*1*), striking the receptive blades of a windmill (*2*), provides an outcome of rotational energy (*3*). In this type of relationship, there is always one impulse activating change, another impulse undergoing change, and a third impulse that is the medium through which the first and second impulses connect and actually produce change. Later in this chapter, we will discover that this triad of *1—3—2* represents the universal action of *interaction*.

In reflecting on giving and taking and the wind's striking the windmill blades, the reader may find it difficult to understand why the relationship should be represented as *1—3—2*. For example, why couldn't the relationship be portrayed as *1—2—3*, since the donor gives to the receiver through the gift, or the wind strikes the blade, which leads to rotation. In fact, the triad *1—2—3* refers to a much different action—what we will call below *expansion*. For the newcomer, the point is that each of the six triads has a particular quality and taste that must be experienced many times before one can gain a thorough understanding of what the six actions are. Of all the systems, the triad is perhaps the most difficult to grasp and to apply. The beginner must have patience and make an effort to apply the triads to situations and events in his or her own experience.[3]

THE MAINLAND, ISLAND, AND BRIDGE

As I have already explained, the three impulses of the triad interact in six different ways (table 3.2). Next, we must understand that each of these triads represents a different mode of relationship that depends on which of the three impulses are the initiating, characterizing, and resultant terms.

To come to a deeper understanding of the six different triads, I first want to describe them through a simple geographical example. Let us suppose that there is a rich mainland with much wealth and close to it an island of meager resources. People of both the mainland and island recognize that perhaps they can help each other, but because of the geographical separation, little is done. In time, however, a bridge is built. The result is a flow of people and goods between mainland and island.

From the viewpoint of systematics, the mainland, island, and bridge can be interpreted as three impulses of a triad. Specifically, the mainland is the *affirming impulse* because it is much more powerful than the island, which is under the mainland's influence and thus the *receptive impulse*. Yet again, the bridge is the *reconciling impulse* between the mainland and island because it provides a link between them. We must ask about the kinds of actions that the mainland, island, and bridge as affirming, receptive, and reconciling impulses can have on each other, remembering that each impulse can occupy any of the three terms of the triad. These various interactions point to the six different triads of table 3.2. I want to examine each of them in turn as they suggest a particular mode of action among the mainland, island, and bridge, which, for convenience, I designate M, I, and B, respectively.

When we consider the possible relationships among mainland, island, and bridge, probably the first action that comes to mind is the mainland's impact on the island. Because the mainland is stronger, its people and goods flow to the island, whose economy and social life are changed. In this triad, the mainland is active and thus the initiating term. The mainland works through the bridge (characterizing term) to interact with the island (the outcome) and eventually to dominate it. This triad is the flow from the mainland over the bridge toward the island and can be represented as M—B—I, or 1—3—2. This triad is the same as the giving-taking and wind-blade relationships above.

At the same time, however, we also must realize that the island takes advantage of what it receives and is able to return goods and services to the mainland. More than likely, the island's people will work diligently because they have fewer natural resources than the mainland. The result is that the island's economic development quickens; contact and commerce increase with the mainland. This relationship is also a triad, but with the island as the initiating term. Because of the bridge, the island gains strength and returns things of value to the mainland. This flow from the island over the bridge to the mainland can be summarized as I—B—M, or 2—3—1.

In these first two triads (*1—3—2* and *2—3—1*), the bridge is the characterizing term. We must also remember, however, that the bridge has come into existence, which means that it can be an *outcome*. On the one hand, the bridge is generated by the powerful mainland, for which the bridge is a vehicle by which the island can become an outlet for the mainland's good and services. In this sense, the mainland is the initiating term, working through the island to create the bridge. This triad can be represented as *M—I—B*, or *1—2—3*. Here, the mainland (initiating term) expands economically through the island as market (characterizing term) with the result that the bridge (outcome) becomes a channel for the greater prosperity and strength of the mainland.

At the same time, the bridge can be the outcome in the triad *I—M—B*, or *2—1—3*. This triad represents the bridge's dependence on the island, which, because of its isolation and indebtedness to the mainland, sees the bridge as a lifeline. In this situation, the island is the initiating term that, through its contact with the mainland, is able to have the bridge. Here, the island (initiating term) puts itself under the economic influence of the mainland (characterizing term) with the result of the beneficial linkage of the bridge (outcome). In one sense, the island's subservience to the mainland allows it to gather itself together economically through the presence of the bridge.

Though the bridge is the outcome of both triads *1—2—3* and *2—1—3*, it is crucial to realize that the actions are greatly different. For the mainland, the bridge as an outcome is a *conduit* for goods and services (*1—2—3*), while for the island, the bridge is a *lifeline* through which the island can improve itself (*2—1—3*). These meanings of the bridge are greatly different, with the mainland emphasizing size and growth, and the island emphasizing permanence and security. The mainland's aim for the bridge is increasing carrying capacity, perhaps even to the point of overload and strain, while the island's aim is continuity and safety.

We must also recognize that the bridge itself is crucial to ties between mainland and island and acts on both of them. In this sense, the bridge can occupy the *initiating term*. Before the bridge, the island lived its own isolated life. After the bridge, the mainland absorbs the island's economy and exerts social and political control. Because of the bridge, the mainland subordinates the island, an action that can be represented as *B—M—I*, or

3—1—2. In this sense, the island's possibilities after the bridge are, from one angle, more limited because now the island is under the domination of the mainland, whereas before it was not. The bridge has brought an order to the island that was not present before.

On the other hand, the bridge as initiating term also leads to the triad *B—I—M*, or *3—2—1*. In this case, the bridge gives the island new opportunities that it did not have before. Islanders are exposed to people and things from the outside. In this sense, the bridge exposes the island to new possibilities and opens it to a new freedom gained with the help of the mainland.

Table 3.3 summarizes the six triads for the mainland, island, and bridge. Note that the triads are arranged in three pairs according to which impulse is the initiating term. First, there are the two actions in which the mainland is the initiating term. Both of these triads involve development and increased strength, either of the island's economy (*1—3—2*) or of the bridge's carrying capacity (*1—2—3*). The emphasis here, however, is *not* on the island or bridge as ends in themselves but, rather, as they are means for the development of the mainland.

Table 3.3   Mainland, Island, Bridge: The Three Pairs of Triads

The Mainland as Initiating Force
M—B—I (1—3—2) Mainland's expanding commerce with the island
M—I—B (1—2—3) Mainland's strengthening of the bridge

The Island as Initiating Force
I—B—M (2—3—1) Island's economic development
I—M—B (2—1—3) Island's concern for the bridge as a lifeline

The Bridge as Initiating Force
B—M—I (3—1—2) Bridge's determining the island's subordination to the mainland
B—I—M (3—2—1) Bridge's releasing new life for the island

In contrast, the two triads in which the island is the initiating term emphasize the well-being and development *of the island itself,* either through its own economy or through the security of the bridge as a lifeline. Finally, there are the two triads where the bridge is the initiating term. These two triads describe the simultaneous actions that the bridge itself brings about: determining how the island is subordinated to the mainland (*3—1—2*) and releasing new life into the island (*3—2—1*).

THE TRIAD OF FATHER, MOTHER, AND CHILD

Before we consider the six triads as universal actions and processes, I want to present one other example—the relationship among father, mother, and child. This triad is less immediately obvious than the above geographical example and is, therefore, more difficult to interpret. In that the father provides the seed for generation, he is the *affirming impulse* (*1*); in that the mother receives the seed and bears the child, she is the *receptive impulse* (*2*); and in that the child is the result of the sexual union between father and mother, he or she is the *reconciling impulse* (*3*).

Remembering that each impulse can occupy any of the three terms in the triad, we have six actions, which Table 3.4 has arranged in three pairs, according to the term that the child represents. We find that these three pairs of triads refer to *generation, family life,* and *individual.*

Table 3.4   The Six Relationships among Father, Mother, Child

The two triads of generation
F—M—C (1—2—3) Father's transmission of seed
M—F—C (2—1—3) Mother's power of attraction

The two triads of family life
F—C—M (1—3—2) The connecting link
M—C—F (2—3—1) The distinctive role of the mother

The two relationships brought about by the child
C—F—M (3—2—1) Mother subordinated to the law of family life
C—M—F (3—1—2) The new life brought by the child

Generation involves "bringing to birth" and is represented by the two triads in which the child is the outcome of the triad. Generation is most readily obvious in the action from father to mother to child (*1—2—3*). Here, the father inseminates the mother, who conceives and bears the child. This progression is mirrored by the complementary action from mother to father to child (*2—1—3*). Here, the feminine attraction of the woman, as a potential mother, acts upon the man, who becomes father of the child. The complementarity of these two triads indicates that the potential of woman for motherhood is as powerful an initiating impulse as the potential of man for fatherhood.

If we consider the two triads in which the child represents the characterizing term, we speak of relationships involving *family life*. Here, the child provides a connection between the father and mother with the result that a family is created. On the one hand, the child's needs are the "ground" through which the father tempers his urges for domination, while the mother sacrifices her own inclinations, with the result that the family can have a clear direction and future (*1—3—2*). On the other hand, the child's needs are the "ground" through which the mother gives strength and a sense of purpose to the father, with the result that the family can have a coherence and stability (*2—3—1*).

Finally, there is the pair of triads where the child represents the initiating term. These triads involve the child as an *individual*. In the triad *3—1—2*, the potential presence of the child leads the woman to require the support of the man to fulfill her task as mother. Because the man is a potential father, the woman leaves her name and home to take the name and home of her husband. In this sense, parenthood imposes on the mother the condition that she subordinate herself to the needs of family life. She comes under the "law" of family life, just as the island falls under the influence of the mainland once the bridge is built. On the other hand, there is the triad *3—2—1*, where the potential child arouses the generative force of the father through the maternal power of the mother.[4]

One difficulty in understanding these six triads of father, mother and child is that the full relationship of marriage and family is much richer than parenthood alone.[5] The nature of human beings is such that they are not only an instrument for the transmission of the human race from generation to generation. In fact, human beings may have other cosmically significant functions to fulfill. Traditional wisdom suggests that there are seven such functions, each of which have a corresponding relationship between man and woman.[6] In addition, children can enter into other kinds of relationships besides generation. The result is a complex network of triads that gives a richness of experience to the relationships of marriage and family.

THE SIX TRIADS

The above examples of marriage and mainland, island, and bridge indicate that interpretation of the six triads requires one to take into account the ways the three impulses interact, depending on their placement as terms. For example, in the mainland-island relationship, the bridge plays six quite different roles:

1. It enables the mainland to act upon the island (*1—3—2*);
2. It enables the island to find itself (*2—3—1*));
3. It experiences growth in that the mainland demands more and more from it (*1—2—3*);
4. It acts as a challenge to the island, requiring that it become stronger to meet the mainland's demands (*2—1—3*);
5. It brings greater order to the island (*3—1—2*);
6. It brings greater freedom to the island (*3—2—1*).

Though it won't be done here, one could also consider the six roles of both mainland and island. The crucial point is that the three impulses of the triad can come together in six different ways that lead to different sorts of relationships and interactions. In fact, these triads describe the six fundamental actions in the world, and I want to describe each of these actions in detail, drawing on the above examples as well as making reference to other contexts. As table 3.5 illustrates, these six fundamental triads can be called **expansion, concentration, interaction, identity, order,** and **freedom.** We look at each in turn—first, in general terms; second, in regard to the triads of mainland-island-bridge and family.[7]

Expansion (1—2—3)

The triad of expansion is associated with involution, creation, growth, generation, and transformation. Expansion involves a movement from unity to diversity. An active agent acts on a responsive ground in such a way that growth results. In this blending of active and passive, the affirming impulse loses some of its original force but gains range and expression. This triad represents any process of creation and, in one sense, answers the question, "How does the universe come into existence?"

In our geographical example, the triad of expansion is represented by the mainland's expanding power through its linkage with the island because of the bridge. In the triad of father, mother, and child, expansion involves generation, whereby the father inseminates the mother to produce a child, who in turn can transmit possibilities to future generations. In this sense, the child is made in the image of the father but is formed in the mother.

Table 3.5　The Six Essential Triads

expansion (1—2—3)
concentration (2—1—3)
order (3—1—2)
interaction (1—3—2)
identity (2—3—1)
freedom (3—2—1)

Concentration (2—1—3)

The triad of concentration is associated with evolution, purification, and unification. In that its direction goes from actual to potential, this triad appears to go against time. Unlike the triad of expansion (*1—2—3*), the triad of concentration cannot proceed automatically. Concentration involves a *purification* in the sense that "the fine is separated from the coarse." Also, the triad involves *unification* in the sense that there is a movement from multiplicity toward unity, a direction opposite that of the triad of expansion. In concentration, the receptive impulse initiates an action toward the affirming impulse, and the result is a new potential. Without the triad of concentration, all order in the world would run down and disperse.

In our geographical example, the triad of concentration involves the island's need to develop by means of the bridge. The island places itself under the action of the mainland to gain greater possibilities. In the example of father, mother, and child, the woman, to become a mother, attracts the man whose paternal powers are aroused, with the child as the result.

Interaction (1 — 3 — 2)

The triad of interaction represents all changes in the world that involve a redistribution of things and energies. Unlike the triads of expansion and concentration, *nothing new is born out of the activity*. The triad of interaction refers to the endless flux of interlocking events continuously going on in the world and being carried out more or less consciously, as with the action of giving and taking, or the action of the wind on the windmill blades.

In our geographical example, the triad of interaction relates to the bridge as a link between the mainland and island; through the bridge, the mainland exerts its actions on the island. In the example of father, mother, and child, the relationship between mother and father can work because of the child's needs, which both father and mother put before their own. Through the initiative of the father, interacting with the mother, family life is given force and direction.

Identity (2 — 3 — 1)

The triad of identity maintains what we already are and what a thing already is. This triad allows all that exists to maintain its own character and answers the question, "How can anything be what it is?" The reconciling impulse in the triad is the inner bond by which everything is what it is. The triad of identity also maintains a thing's place and context in the world. The fact that all things have a place gives them a power to assert themselves.

In the geographical example, the island is connected by the bridge with the mainland; through this union, the island gains strength and greater identity. In the example of father, mother, and child, the mother can demand constancy from the father through the medium of the child; the result is that the family gains stability and greater identity.

Order (3 — 1 — 2)

The triad of order can also be called the triad of determination or the triad of consistency. This triad maintains universal order and relates to the question, "Why is it that everything has to be as it is?" The triad of order explains why the world cannot be capricious and arbitrary. Rather, things must always be as they are, thus the future arises from the past, but the past can never arise from the future; nor can water run uphill or the earth move about the moon. Constancy is integral to the world and represented by the triad of order.

In the geographical example, the triad of order works through the bridge, by which the island is brought under the economic and social life of the mainland.[8] In the example of father, mother, and child, the mother, through the child, falls under the influence of "family," whose needs she actualizes through the father.

Freedom (3 — 2 — 1)

The triad of freedom helps explain why the world can be otherwise than it presently is. This triad involves the sense that there is something in us which is free, and that this freedom is as important as that of universal order. In the triad of freedom, the reconciling impulse as initiating term can be associated with an *opening* through which possibilities happen that otherwise could not. A creative impulse is liberated in the world because something new seeks to be born.

In the geographical example, the bridge as initiating term brings about new life for the island. The bridge provides the island with possibilities that could never be available otherwise. The island gathers new energy that directs it to new enterprises and, in one sense, sets it free. In the example of father, mother, and child, the child's wish to be born touches the mother who draws the father to her. The child as initiating term creates the possibility of a father through the mother.

THE SIX TRIADS IN RELATION TO HUMAN SOCIETY

As a final example of the six triads, I want to consider the monad of human society, which as a triad can be said to involve the three impulses of *rulers*, *ruled*, and *reformers*.[9] Clearly, the affirming impulse of human society is the authority of the few, or the rulers; the receptive impulse is the passivity of the many, or the ruled; and the reconciling impulse is those individuals of society who are neither rulers nor ruled but, instead, seek to bring about change—that is, the reformers. On the one hand, reformers may question the status quo and reflect anti-establishment sentiments. On the other hand, they may be the base for creative societal change in the form of inventions, ideas, practical solutions, peace making, creative expressions, and so forth.

In all societies, whether democratic or oligarchic, modern or traditional, capitalist or socialist, there is a relationship among the few who have the power, the many who are under that power, and an independent element that in some way seeks change. These reformers do not always set themselves consciously to reform their society, but changes occur none the less.

When we study the dynamics of society in terms of these three impulses, we arrive at the six different triads, which, in this case, can be called *creativity, concentration, interaction, conservation, justice,* and *freedom.* We assume in all these triads, that the "ruled" remain receptive, that the "rulers" remain active, and that the "reformers" link the ruled and rulers together. We must not forget, however, that these three impulses can occupy different terms in the triad.

The triad of *creativity* exemplifies the more universal triad of expansion (*1—2—3*). The action of the rulers corresponds to the needs of the ruled, and the society flourishes and adapts to changing circumstances. The heart of this triad is *transmission,* and the "reformers" of one phase become the "rulers" in the next. An example is Plato's conception of education, where the younger generations are the bearers of social progress. In simple form, the triad can be expressed: The authority residing in the rulers adapts itself to the needs of the ruled to stimulate the reformers' creative activity by which the society grows.

In the second relationship, the triad of *concentration* (*2—1—3*), changes in the social order arise from the needs and pressures of the ruled. One example is the English reform movement of the 1850s where the "reformers" appeared in the establishment of the British Civil Service. Unless Great Britain's modern Civil Service undergoes a fresh triad of concentration, it will degenerate into a useless bureaucracy, as was the case in Russia after the liberation of the serfs.

The simplest form of the triad of concentration is a direct appeal to a ruler, as with the case of the brilliant thirteenth-century Russian administrator Alexander Nevsky, who submitted to the demands of the Mongol overlords and so averted their continual devastating attacks. The triad of concentration can be expressed: The needs of the ruled evoke from the rulers the readiness to sacrifice some of their authority to enable new elements to make their contribution to social progress.

The third relationship is represented by the triad of *interaction* (*1—3—2*) where "reformers" refer to that portion of the society which keeps it in a state of progress. Generally, this group of people is some middle class that activates and maintains the activities of the society. For example, Plato spoke of such a middle class when he described "workers for the people," which he regarded as the heart of a well-governed city. Throughout history, a middle class of merchants, craftspeople, and managers has almost always provided the practical connections between the rulers and the masses, particularly in terms of the economic dynamism of the society. "Higher" modes of interaction are represented by the professional classes, including specialists and administrators. The triad of interaction can be expressed as: The rulers sponsor activity that is regulated by the middle classes and produced by the ruled.

The fourth relationship involves the triad of identity (*2—3—1*), or *conservation*, as I call it here. In this triad, the middle class not only provides opportunities of exchange between the rulers and the ruled but also provides a stabilizing element that gives the society its distinctive character. The "reforming" function here expresses itself through public service in education, medicine, art, science, administration, and so forth. This triad can be expressed as: The identity of a society is founded upon the majority, shaped by the middle class, and made visible by the rulers.

The fifth relationship relates to the triad of order and is called the triad of *justice* (*3—1— 2*) because it involves societal fairness. A stable society requires authority without tyranny. A judiciary must be available to adjust the balance between rulers and ruled. This judicial function is not only confined to official magistrates and formal law. Public opinion, the *consensus omnium gentium*, illustrates the power of the reconciling impulse to see that responsibility tempers authority. This triad can be expressed as: The social conscience manifested by those not directly involved in ruling stabilizes the relationship of ruler and ruled.

The last relationship involves the triad of *freedom* (*3—2—1*). Here, the reformers come into their own as champions of freedom. By serving the interests of the masses and initiating reforms, the reformers enable the authorities to act creatively on the society as a whole. This triad can be expressed: Creative freedom present in some people invigorates the life of the majority and enables the rulers, as instruments of the larger society, to exercise authority.

MISCONCEPTIONS

In working with the six triads, it is easy to go astray, and I want to point out some potential problems and confusions in regard to the six triads of human society. Perhaps this discussion will assist students as they consider other actions and relationships as triads.

A first common mistake involves the distinction between the character of an impulse and the particular term it occupies in the triad. For example, one often confuses the affirmative character of the rulers with the notion of initiative; one supposes that, within the same relationship, the affirmative character can pass from the few to the many. Consider the triad of concentration (*2—1—3*) as an example. Often, students ascribe revolution or the "assuming of power" by the masses to this triad. If one looks at history carefully, however, he or she realizes that mass movements and revolutions are *always the work of the few*, despite slogans and theories to the contrary.

In these situations, the original rulers representing the affirming impulse are displaced by a newer group of rulers. Whoever the particular rulers, *the masses remain passive*. At the same time, however, we must remember that the ruled, though always representing the receptive impulse, *can* occupy the initiating term. Here, it is exactly their receptive quality that evokes one of two actions: in the triad of concentration (*2—1—3*), the needs of the ruled stimulate the rulers to make constructive changes with help from the reformers; in the triad of conservation (*2—3—1*), the reformers adjust to the needs of the ruled directly and stimulate stabilization.

This example illustrates the crucial importance of distinguishing the character of an impulse from its role as a term. If we don't understand this difference, then we will always assume that the initiative is in the hands of the affirming impulse. In regard to the family, for example, one would assume that the initiative is always with the father. But this assumption is true in only two of the family relationships—the triads of expansion (*1—2—3*) and interaction (*1— 3—2*). The mother and child make equally important and necessary initiatives, though they continue to represent their respective receptive and reconciling roles.

One may also wonder what triad is at work if the rulers do not rule effectively and the reformers must become involved. This situation is described by the triad of interaction (*1—3—2*), but beginners often interpret the situation erroneously as the triad of expansion (*1—2—3*). This latter triad can only arise successfully when there is a perfect correspondence between the affirming and receptive impulses. Otherwise, they cannot work on each other. If this correspondence does fail (as, for example, in misrule), some intervention is required. If this intervention is the reconciling impulse, then that impulse is consumed in reestablishing a correspondence; the impulse is not free to transmit its virtue or to initiate new possibilities.

Another point that students often forget is that all six triads are always present in every situation. Theoretically, each of the three impulses is sufficiently true to itself and, therefore, has the power to occupy any of the three terms. In this case, all the triads are present and the situation is harmonious. In the real world of human societies, however, this situation can never happen because the world is not ideal. The best way to judge a particular human society is to examine it in terms of the six triads and to consider which have broken down and why.

Another question that sometimes comes to the beginner's mind is why, in an ideal society, would the reconciling force not be transmitted by the rulers? This situation is impossible because the ideal society involves a threefold nature that includes (*1*) those few committed to the exercise of authority, (*2*) the many who depend upon authority, and (*3*) those who are neither dependent nor committed, who neither exercise authority nor depend on it. Only the last group—the reformers—are free to make adjustments.[10]

Another question students often ask when considering the triads of human society is what triad a dictatorship or a democracy is. Both forms of society involves *all* the triads. For example, if a dictator accepts independent advice rather than relying only upon his own judgement, then all the triads are possible. If he rejects independent advice, however, he eventually fails and his regime falls.

The difference between a democracy and a tyranny is in the way in which adjustments are made. In other words, what degree of freedom is there for the particular society to adapt itself to changing conditions? In the case of a tyranny, the dictator can be overthrown; in the case of a highly organized and structured society like the modern welfare state, it may be much more difficult to produce changes.

THE SIX TRIADS AND SELF-DEVELOPMENT

If the six triads represent universal processes and actions, they should have relevance to our own worlds of experience and personal development. In this section, I discuss the significance of the six triads in regard to self-growth and personal transformation.

Perhaps the most important of the six triads for learning and self-development is *concentration* (*2—1—3*). Here, the affirming impulse is *not* the initiating term, a fact that may at first seem impossible, since we generally suppose that we only improve ourselves through our own power and desire. Such an active relationship to the world, however, is only one part of human experience. More thorough personal development only occurs when we put ourselves in front of a challenge—that is, when in a receptive mode, we face a situation and thereby place ourselves in the position of the affirming term. One example is Arnold Toynbee's *A Study of History*, which argues that great civilizations only arise when they confront some challenge; conversely, they fall when the challenge disappears.[11] We can also recognize this pattern in our own lives: if we stand up to a challenge and succeed, something in us grows stronger.

The triad of concentration helps us to understand that this "something" is equivalent to the bridge in the mainland-island example. I have some inadequacy. There is something I wish to do but cannot. I begin to train and undergo hardships to acquire the skills and strength to overcome the inadequacy. As a result, something gradually changes in me. This "something" is a kind of bridge that integrates me inwardly and makes me stronger in the way I am connected in myself.

This action *concentrates* my inner unity and being. Where there was less inner connectedness, there is now more. I feel a strength and freedom that are only present because I have faced the challenge successfully. In this confrontation, my weakness is not changed. Rather, the triad of concentration creates a bridge that draws together what before was unconnected within me. When we experience this triad, we feel the presence of a new inner life—an awareness not of our strength or weakness but of our inward freedom.

Very much different from the triad of concentration is the triad of *identity* (2—3—1), which involves the repetition of an action to the point that we absorb it completely, and it becomes an integral part of who we are. The repetitive quality of this triad fosters a strength of affirmation, which is much different from the sense of freedom arising from the triad of concentration. In the triad of identity, the bridge, or connecting factor (3), is the continuous repetition of the same act or skill—the many hours required in learning to write or to sculpt or to do woodworking, for example. With practice and diligence, one masters the particular action, which becomes a part of who he or she is as a person.

The potential danger of the triad of identity is that, if the person continues to repeat only one action, he or she may lose the power to do anything else. One imagines the mechanic or intellectual or artist or housewife who can do nothing but his or her own work. Although repetition can make a person stronger in the action being repeated, it does not enable him or her *to become free.* The person seeking to understand self-development must learn well the triads of concentration and identity. What does one obtain from confronting a challenge, and what does one get through a particular repetitive action?

The triads of concentration and identity both have the receptive impulse as the initiating term. We turn now to the triads of interaction and expansion, both of which have the affirming impulse as the initiating term. Like the triad of identity (2—3—1), the triad of *interaction* (1—3—2) can be taken to represent ordinary life. In terms of self-development, there is nothing special about this triad, since it reflects the way that all people live their lives. In this triad, people act and are acted upon by their environment; they receive external influences, build up a certain experience (in part through the triad of identity), and become a certain kind of person as a result. The triad of interaction reflects the ordinary process of living.

The triad *1-2-3* is the triad of *expansion*. Here, the affirming impulse works through the receptive impulse to create some reconciling impulse as outcome. In terms of self-growth, this triad is crucial and involves situations where we put ourselves in the face of our own weaknesses with the result that we gain freedom. One example is an individual's placing himself under the influence of a teacher. The teacher (*1*) takes the initiative, and the student (*2*) obeys. This action can lead to constructive changes in the student, *provided the teacher knows exactly what the student needs.* In the triad of expansion, no adjustments are possible; there must be an exact fit between affirming and reconciling impulses. The student-teacher relationship, therefore, involves considerable risk, since the teacher's affirmation may not correspond to the needs of the student.

This exact fit is characteristic of both triads of expansion and concentration. Both triads involve a direct connection between affirming and receptive impulses. For the triad of concentration (*2—1—3*), the risk is considerably less, since the initiative comes from the individual who brings his or her own weakness in front of a challenge. In the triad of expansion (*1—2—3*), in contrast, the individual gives the initiative to another person or some other force outside the individual's own control. The contact must be exactly right; otherwise, the outcome may not correspond to the individual's nature and becomes an unhelpful imposition.

To clarify this point further, we can return to the example of the mainland, island, and bridge. If there is no real commonality of interests between the mainland and island, then the mainland will sooner or later impose its conditions on the island, since the mainland is stronger. Whenever we submit ourselves to the action of expansion, we take this risk, since the triad depends on a powerful force as initiating impulse. Once one is under this force, he or she can no longer adjust the situation in the way that one can with the triads where the bridge works as a moderator of flow between the mainland and island—that is, in the triads of interaction (*1—3— 2*) and identity (*2—3—1*).

In both the triads of concentration and expansion, the bridge, or inner connection, is the *outcome* of the action. On the one hand, this outcome can be damaging if affirmative and receptive impulses are not complementary. On the other hand, if the connection is harmonious, the affirmative impulse gives birth to freedom or new life. For example, human pro-creation involves the triad of expansion. The sexual union between man and woman gives birth to a new human being This particular example of expansion generally involves no risk, since it involves the natural process of generation. When the triad of expansion involves psychological and spiritual development, however, the laws of nature are no longer suffi-cient and the dangers of error and failure become much greater.

The triads of expansion and concentration are the basis of psychologi-cal and spiritual *transformation*. When applied to individual develop-ment, their outcomes, represented by the reconciling impulse, involve an inner union within the person himself. These actions must be right if per-sonal transformation is to happen successfully. In this sense, these triads involve *hazard*. There is no freedom of adaptation as there is for triads where the third force is in the middle position. Because of the hazard as-sociated with the triads of expansion and concentration, spiritual tradi-tions always include safeguards such as responsible teachers, recognized practices, and actions that by their very nature produce the right kind of relationship.

The last two triads we need to discuss in terms of self-development are *order (3—1—2)* and *freedom (3—2—1)*, both of which have their recon-ciling impulse as the affirming term. The bridge as an affirming term in the geographical example was deceptively simple, and to understand these two triads in regard to personal development is much more difficult. It is crucial to understand in all the triads that the reconciling impulse is al-ways mysterious, especially in regard to our inner life. Gurdjieff said that human beings, by their very nature, are blind to the reconciling impulse, or "third force," as he called it. We can readily understand affirmation and experience it as a desire or an urge to do. In the same way, we can readily understand receptivity and the experiences of passivity, responsive-ness—even opposition and denial. We can much less readily understand the third state of reconciliation. We may think we can know it, but in truth we cannot.

The more one studies and understands the six triads, the more he or she realizes that the third force is hidden and miraculous. We only par-tially see it manifested through something. In the triad of father, mother, and child, for example, we can only partially grasp the third force; behind the child is a much more mysterious process. In the geographical example, I chose a bridge because it gives some impression of the third force as a connection. But this example is artificial and in many ways deceptive. When we apply the triads of order and freedom to people's inner life, the mystery of the third force is inescapable. We do not know what it is that produces the inner connections.

This impossibility of knowing explains why the triads of order and freedom, both initiated by the reconciling impulse, are the most mysterious of all actions. These two triads answer two questions that have profound metaphysical significance: Why is it that everything has to be as it is? And why is it possible that things can become different? Let us consider the first question. Why, for example, is it that today must come after yesterday, or that two and two cannot make five? Why is it that the laws of gravitation and movement and life necessarily exist? We come to see that the world exists in the only way it can, yet no one can know why these patterns are exactly as they are.

These questions and observations point toward the fifth universal action—the triad of *order* (*3—1—2*). Through this triad, *all things are as they are*. The world is not capricious and arbitrary but full of necessary and inescapable pattern. As long as we live, we are certain that today will never return to yesterday but will always go ahead to tomorrow. We are certain that we will never see any universal laws violated or upset. We take this universal order for granted and assume without question that it is a part of our own lives. The triad of order says that we are what we are, and this fact is not capricious or arbitrary. Things are kept in their own place as what they are, and the world obeys its own laws.

Paradoxically, we also believe that within this firm, universal order there is also *freedom*, which points to the sixth universal action—the triad *3—2—1*, which has the affirmative impulse as an outcome. The triad of order does not deny the possibility of freedom. The world must be as it is, but it also can be otherwise. Both being and becoming are possible in the world because of these two triads.

In regard to the example of the mainland, island, and bridge, the triad of order is reflected in the way that the bridge determines the patterns of connection between the mainland and island. These connections can only be what they are because the bridge is what it is. In regard to the triad of freedom, the bridge opens possibilities that were not present before. In the end, however, these explanations of the bridge are artificial and inadequate. The bridge symbolizes a mystery that can never be understood completely. This mystery is the incomprehensibility of the world.

From one point of view, the triad of freedom illustrates the way that God works in the world from the outside. The affirmative power as outcome is transmitted through all existence and can be imagined as a creative power that branches out through everything, becoming weaker and weaker. Eventually, this force reaches humankind and then passes on to animals, plants and material things, all the time growing fainter and fainter. This creative impulse penetrates everything, but it begins very far away. By the time it reaches us, it is something quite different—little more than a power of affirmation or assertion.

In the same way, there is also a universal existence represented by the triad of order. In its most obvious manifestation, this order is matter and the material world. But we human beings, too, must accept being acted upon. Like freedom, this action also comes from God. The whole of creation is the result of a creative power, but this power only enters the world directly as the reconciling impulse in the initiating term. As I've already said, the third force is mysterious. We can never catch it because it is universal and in everything and it can adapt itself to everything, taking an infinite variety of forms.

On the one hand, the reconciling impulse makes certain that the whole world obeys the purpose of the Creator. On the other hand, the reconciling impulse makes certain that the whole world is free to fulfill itself in its own way. This latter possibility is especially true for us human beings: we are given freedom and responsibility for our own destiny. It is as if God lent us a part of himself to do with as we like. This part is not merely something that he has made; this part is joined and arises from his own very nature that consists of complete freedom.

In a true sense, the reconciling impulse as initiating term is a bridge between Creator and creation. This bridge is intimate and enters into everything everywhere. As it approaches us men and women, this bridge has the special quality that allows us to become free and independent. This quality is not given without payment, however. To touch this higher freedom, we must create our own reconciling impulse through efforts and diligence. Here, we return to the triads of concentration and expansion, where the reconciling impulse becomes the outcome and we create our own third force.

FINAL COMMENTS ON THE TRIAD

In working with triads, we must first be clear about the nature of the structure being studied. We must, in other words, first understand the structure as a *monad* with its particular content and limits. Second, we must examine the structure as a dyad, particularly looking for tensions and complementarities that might provide an opening for the reconciling impulse.

In looking at the structure from the vantage point of the triad, we should first remember that the three impulses must allow in their natures for dynamism and interaction. When we use a thing, quality, activity or some other element to represent an impulse, we must be sure that we interpret the element not as a thing unto itself but *as it represents the impulse*. Sometimes, this process is facilitated by associating the three impulses with the following characteristics:

affirming impulse: pressure, domination, force, or urge;
receptive impulse: potentiality, vehicle, capacity, ground, denial, or resistance;
reconciling impulse: life, transformation, freedom from form.

Only by identifying all three impulses, can we recognize the particular action. In other words, we can only see the meaning of an impulse through the full triad. Ideally, the triad underlying the particular action or situation should arise intuitively. Rather than figure the triad out intellectually, we should find a way to "taste" it—to sense it as a whole first. Then, as far as possible, we work to describe the particular triad in one or a few words. Only then should we elaborate the triad's specific dynamics. Finally, if we speak of one triad, we should consider how this understanding relates to the other triads of the structure we are studying. We should consider each action in relation to all the six triads as they are taken together and are present in the monad.

NOTES

1. See John G. Bennett, *The Dramatic Universe: Volume 2, The Foundations of Moral Philosophy* (London: Hodder and Stoughton, 1961), part 11; *The Dramatic Universe: Volume 3, Man and His Nature* (London: Hodder and Stoughton, 1963), pp. 23-29.

2. G. I. Gurdjieff, *All and Everything: Beelzebub's Tales to His Grandson* (New York: Harcourt Brace, 1950).

3. In volume 2 of *The Dramatic Universe* (p. 100), Bennett suggests why the triad is more complicated than other systems: "We shall find that the property of an endless variability is inherent in the triad and in no other system. This is due to its special position in the series, where it stands between the polarized fixity of the dyad and the concreteness of the tetrad."

4. For further discussion of the six triads of father, mother and child, see John G. Bennett, *Sex* (York, Maine: Weiser, 1989), chapters 4, 6, and appendix II.

5. For further discussion, see John G. Bennett, *Sex*, note 4.

6. Ibid.

7. In Bennett's original lectures, the six universal triads are described in skeletal fashion. In each of the six sections that follow, the editor has written an introductory paragraph that provides a general sense of the particular triad. These paragraphs are based on portions of Bennett's extended discussion in volume 2 of *The Dramatic Universe*, chapter 28, "The Six Fundamental Laws." Readers who wish to understand the six triads more deeply should study part 11 of volume 2, "The Triad—Will."

8. "This growing control, however, does not mean that the mainland absorbs the island completely. If this were the case, the bridge would no longer be an independent term and there would no longer be a triad."

9. Both this section and the following are not part of Bennett's original lecture on the triad. Instead, the editor has assembled these sections from the answers that Bennett gave to a question he asked participants at the end of the lecture: "Would you see if you can work out the six basic triads as they might apply to human society? Assume the following three impulses: the oligarchy, or *the few that have power*; the majority, or *the many who are passive*; and what can be called the reformers, or *those who bring about various creative changes in society*, though not always intentionally."

10. "One example is the holy man Ashiata Shiemash in Gurdjieff's *Beelzebub's Tales*. In Ashiata Shiemash's reorganization of society, the authorities no longer exercise power for the purpose of domination."

11. Arnold S. Toynbee, *A Study in History*, 12 volumes (London: Oxford University Press, 1934-1954).

CHAPTER 4

The Tetrad

THE FOURTH-ORDER SYSTEM is the *tetrad*, which relates to activity and answers the question "What is happening here and why?" This question addresses any situation whereby different elements have somehow to blend and to act upon one another. Cooking, for example, involves an activity in which ingredients are cut, mixed, heated, and cooled. To understand cooking as an activity, one must ask questions such as: "Who is the cook?" "What is the meal to be and what recipe is used?" "What are the ingredients and what is being done with them?" "What cooking facilities are available?" "How will the meal be served?"

From the vantage point of systematics, an activity like cooking can be understood in terms of the tetrad. The tetrad has four terms, which are called **sources**. To understand these four sources and the way they mark out activity, we must first examine the purpose of an activity and the knowledge available for bringing it about. In addition, we must understand the base from which an activity begins and the instruments and abilities available for its accomplishment.

For example, let us return to the activity of cooking. Here, the purpose is a meal and the knowledge for preparing that meal involves a chef. The ground for cooking is the raw ingredients, and the instruments and abilities are the kitchen facilities and the chef's skills. In every activity there are elements that correspond more or less to these four dimensions or sources, though the way they enter into human experience is infinitely varied. These four sources can be symbolized as the four end points made by a cross of vertical and horizontal lines (figure 4.1).

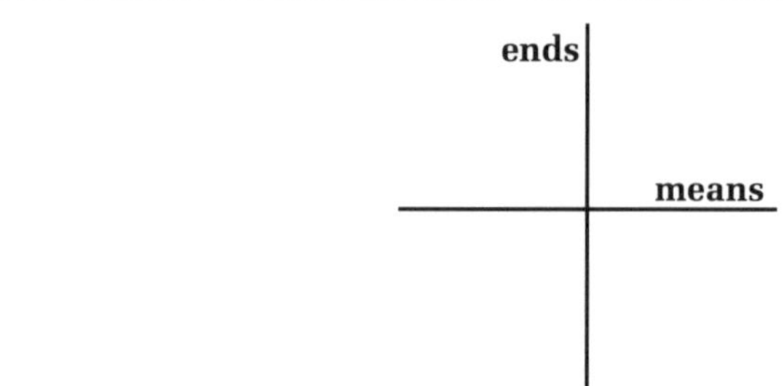

Figure 4.1   The Tetrad as Means and Ends

NAMING THE FOUR SOURCES

Each activity in the world is different. This diversity of expression poses a difficulty in finding appropriate names for the four sources. To help to establish labels, we can first consider the vertical and horizontal lines of figure 4.1. The vertical line represents what might be called the *ends*. Roughly speaking, we can say that the bottom point of this line represents the raw material, or the *ground* for the activity, while the top point represents the finished product, or the *result* of the activity.

These terms are general and do not adequately describe all processes. A key aspect of the vertical line is *quality*. To cook a meal, for example, is not simply to change the character of the raw ingredients. In addition, one works to prepare the dish perfectly so that the meal will give enjoyment. This quality of excellence and pleasure represents the highest point of the vertical line. In one sense, the bottom point of the vertical line can be called "quantity" and the top point, "quality." One can say that the vertical line marks the effort to transform quantity into quality. This distinction, however, does not accurately describe all activities, just as the labels "ground," "raw materials," "result," and "finished product" are not accurate either. Here, I use the terms "spiritual source" and "material source" for the top and bottom points of the vertical line.

The **spiritual source** is the goal or end of the activity. This source draws the process to completion and indicates the direction where significance and quality are found. In this sense, the spiritual source represents the place of *value* in the activity. The spiritual source also represents the *ideal* and points to the unseen and the supernatural. In this sense, the spiritual source is beyond form. The **material source**, in contrast, designates the starting point of raw material and relates to such qualities as ground, fact, quantity, and actuality. In this sense, the material source is *natural* and involves an extended *spatial* nature. Whereas the spiritual source represents the ideal pattern of an activity, the natural source represents the "stuff" out of which the pattern is wrought.

We also must consider the end points of the horizontal line and provide them with source names. One of these sources provides the practical means whereby the process can go forward, while the other source motivates and sustains the process. This set of sources might be called "instrument" and "agent" or "practical" and "theoretical." In the case of cooking, the chef and equipment would represent these terms. The chef knows both the recipe and the techniques for preparing it, but this preparation is only possible with utensils, appliances, fuel, and so forth.

We might also interpret the horizontal line in terms of theory and practice, since these labels provide an appropriate terminology for a great range of human enterprise, from bodily to intellectual activities. Still, these labels are not sufficiently general and complete to give the depth of meaning equivalent to the distinction between material and spiritual.

Here, I use the terms "masculine" and "feminine" to identify the sources marked by the end points of the horizontal line. Because of their association with gender, these labels at first glance may seem arbitrary and inappropriate, but one must understand that I use these terms symbolically. The **masculine source** refers to a driving quality concerned more with the theoretical side of the activity. In this sense, the masculine source represents knowledge and thinking and points toward the creative awareness directing the activity. In contrast, the **feminine source** refers to a quality that works to hold an activity together and to give it form practically.

Again, let me emphasize that I use the words "masculine" and "feminine" in a symbolic way. It would be entirely incorrect to conclude that all men somehow occupy the masculine source and all women, the feminine source. There is in everyone, men and women alike, a certain masculine and feminine power that has nothing to do with the gender distinctions of male and female. Instead, these powers relate to the forces of activity that flow in us. It is obvious that men can be as skillful and practical as women, just as women can be as thoughtful and as knowledgeable as men.

In the end, "masculine" and "feminine," as well as "spiritual" and "material", are too narrow to suggest the full wealth of meaning symbolized by the four sources. One must work continuously to keep a broader sense of the four terms' significance in mind. Figure 4.2 represents the four sources graphically and provides elaboration of the simple ends-means distinction illustrated in figure 4.1. Note that the four sources have been connected with lines, which signify additional important links among the sources. Shortly, we will consider these links in greater detail.

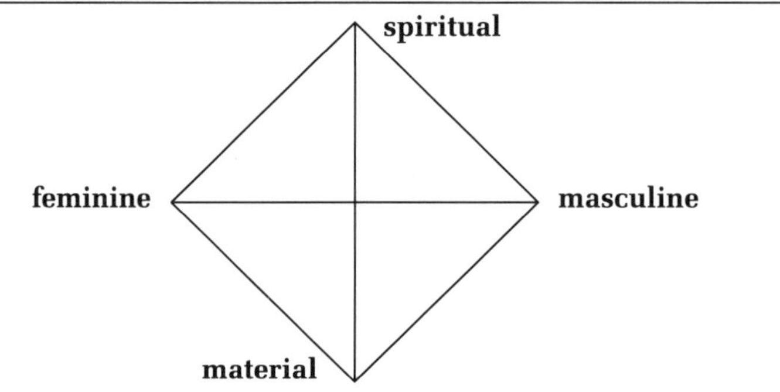

Figure 4.2   The Four Sources of the Tetrad

The notion of the four sources is one of the most ancient symbolic traditions. For example, the four sources appear in the Book of Genesis in a very strange way, in the four verses of the second chapter describing the Four Rivers of the Garden of Eden. According to many Biblical commentators, these verses have no place here, for they are out of character stylistically with the rest of the chapters. Whether out of place or different because they were to be given special emphasis, these verses describe the idea of four rivers arising from one river (though it is unclear whether the rivers flow into or out of the Garden).

In more recent times, the four sources have appeared in other guises. As I pointed out in the introduction, Carl Jung described the fourfold nature of human beings through the four sources of intuition, sensation, feeling and thought. In Jung's tetrad, intuition is the higher human power that touches spiritual realities, whereas sensation relates to the material source. Yet again, feeling and thought represent the feminine and masculine sources. Figure 4.3 illustrates Jung's fourfold scheme as a tetrad.

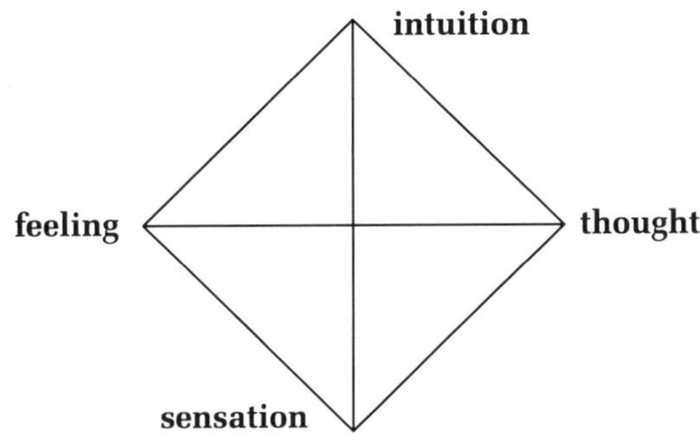

Figure 4.3 Jung's Fourfold Typology

SCIENTIFIC ACTIVITY AS A TETRAD

To illustrate the use of the tetrad in understanding activity, I want to examine modern western science, which provides one means to gain deeper insight into the nature of the world. If the tetrad accurately describes activity, we should be able to point to four sources that delineate scientific activity and help to elucidate degrees of success for science.

One of the first essential requirements for scientific endeavor is *accurate contact with the thing being studied.* Modern western science really began when the idea that one could study nature without direct encounter was replaced with a requirement for firsthand contact. This empirical immediacy is the starting point of scientific activity and corresponds in the tetrad to the material source—the actual, visible facts with which the scientist must deal.

If one studies many of the great advances in western science, one realizes that they have often involved direct insight into phenomena. It is uncanny how a scientist like Lord Rutherford or Faraday could, without invoking any general principles, understand a particular process and create experiments that could never be imagined if approached only through conceptual knowledge or technical skill. Intellectual and practical qualities are important but, for scientific activity, they cannot replace direct contact with the phenomenon.

If, however, the scientist masters only the material source and has no greater unifying vision, his or her work will remain small and never come to life. Significant scientific activity is marked by *a special kind of wonder and faith*. The scientist must have insight, vision, and a sense of nature's mystery. As Isaac Newton said, "I do not know what I may appear to the world, but to myself I seem to have been only like a boy playing on the seashore, diverting myself in now and then finding a smoother pebble or prettier shell than ordinary, whilst the great ocean of truth lay all undiscovered before me."

Newton's sense of wonder for nature relates to the top term of the tetrad—the spiritual source. This term of scientific activity does not involve a blind worship of nature but, rather, touches upon a sense of underlying pattern. The spiritual source represents the ultimate goal of scientific activity: to move continuously closer to a comprehensive understanding of the world. For scientific activity, the spiritual source might be called *vision*.

Actualization of vision involves the two vertical sources. The masculine source can be identified as *knowledge*, which links the scientist with past and future. The masculine source involves intellectual comprehensiveness and the ability to relate one's own ideas and research to the larger discipline of which the scientist is a part. The masculine source points to the requirement that a scientist must be erudite. Researchers who do not know their field thoroughly may mistake the significance of their research, relate it improperly to other discoveries, or conduct work already accomplished or disproved by other scientists.

At the same time, the scientist must be able to conduct research practically, and this need relates to the feminine source of the tetrad—what might be called skill or *technique*. In this regard, the scientist must be able to conduct effective experiments and to observe the phenomenon clearly and completely. Technique is exemplified, for instance, by the persistence and courage of Madame Curie, who conducted a long series of difficult experiments that only slowly showed results.

The full tetrad of scientific activity is illustrated in figure 4.4. One intriguing exercise is to use this tetrad to assess the careers of great scientists. No scientist's work is so perfectly balanced that all four sources play an equal role. Some scientists have a knack for seeing empirically, while others have the ability to synthesize research in a field and to integrate their own work accordingly. Yet again, some scientists have great technical skill and a determined persistence to carry their work through, while others are visionaries who can see deeply into the principles of nature. Einstein, for example, conducted theoretical experiments on paper and had little interest in empirical research or practical techniques. He had a remarkable ability to integrate scientific knowledge and to see conceptual patterns hidden from other scientists.

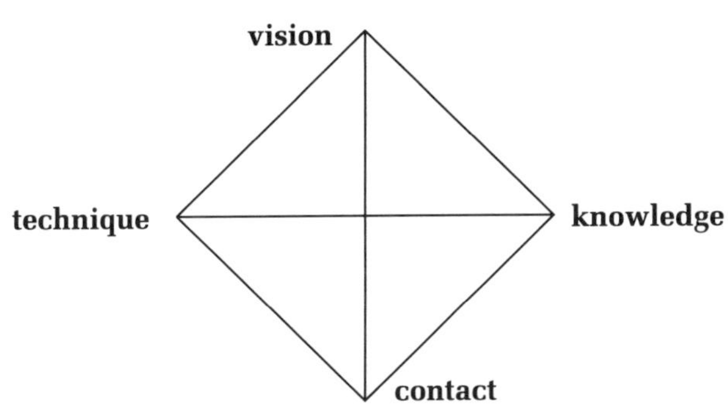

Figure 4.4   The Tetrad of Scientific Activity

CONNECTIONS AMONG THE FOUR SOURCES

The example of science illustrates how the tetrad can be used to examine a particular activity. Next, it is important to understand that the four sources of the tetrad are not separate but *interconnected and influencing each other*. A complete understanding of a particular activity is possible only if one examines how the four sources interrelate.

If one studies the tetrad, he or she notes that there are six connections: the original two lines making a cross plus the four lines making the outline of a diamond. As we've already seen, the vertical line connects the material and spiritual sources. This line represents the entire range of motives driving the particular activity forward. These motives often seem irrational and beyond control, since they cannot usually be explained logically.

Likewise, the material source is often beyond reason and control because it is what we are given. The actual situation must be accepted as it is, since little can be done to change it. Similarly, the spiritual source is beyond reason and control. This source is the ideal that the activity seeks, and it *is what it is*. We seek this aim but we cannot make it happen. It can only unfold in its own way and time.

In suggesting that the vertical line and its sources are largely beyond control, I do not mean to suggest that nothing can be done by the individuals involved in an activity. Obviously, a biologist may suddenly decide to study theoretical physics, and in this sense he changes the material source of his work. This shift, however, cannot be done easily, since he cannot invent what he wants but must take the facts that are given to him by the new field. At the same time, we must suppose that all scientific activity, whether biology or theoretical physics, eventually meets at some ultimate top point on the tetrad—what we might call *truth* or *reality*.

Any point along the vertical line represents an intermediate position between material and spirit, fact and value. In other words, the motivation for any activity is never necessarily pure and can involve all shades of driving forces. Any real activity involves a complicated set of motivations; it is never so simple as brute facts, on one hand, and a certain ideal to achieve, on the other. Preparing a meal, for example, can never be reduced down to available food and workable recipe. The chef must also consider factors of time, facilities, and personnel. The motivations for the meal are complicated by all of these factors. There is no guarantee that right materials and workable recipe will lead to the "perfect meal" that the chef envisions.

We now turn to the horizontal line and its sources. They reflect aspects of the activity that *can* be controlled, at least to some degree. Through effort and practice, we can develop our knowledge and skills. In this sense, one hones the means of the activity and provides a better vehicle to move toward the spiritual source. On the horizontal line appears the value of repetition and experience: that by participating in an activity many times, one becomes more skilled and knowledgeable.

As a whole, the tetrad represents action involved in aim. The ideal action is from material to spirit and back again. But this reciprocal movement has no actualizing body, thus we come to the vertical line, which suggests possibilities for the *what* (masculine source) and the *how* (feminine source) of the activity. Theory and practice and thought and feeling come together to express the particular activity.

With the complexity of the vertical and horizontal lines clear, we can now summarize the six connections among the four sources of the tetrad illustrated in figure 4.5:

1. The vertical line between physical and spiritual sources connects fact and value. Fact is realized by acquiring value, and value is made real by possessing fact. This relationship can be called the **line of realization**.
2. The horizontal line between masculine and feminine sources joins the cognitive and instrumental aspects of the activity. Realization is given substance through the vehicles of knowledge and technique. The result is a mutual completion that can be called the **line of understanding**.
3. The upper right line between spiritual and masculine sources connects value with the cognitive aspects of the activity. The effort is made to unite knowledge with value, which leads to the embodiment of value in form. This relationship can be called the **line of participation**.
4. The upper left line between spiritual and feminine sources joins value with the instrumental aspects of activity. The practical instrument accepts the operation of value and thereby participates in a non-material process. This relationship can be called the **line of transformation**.
5. The lower right line between material and masculine sources connects fact with the cognitive aspects of activity. Sensation turns into knowledge. This relationship can be called the **line of perception**.
6. The lower left line between material and feminine sources joins fact and instrumental aspects of activity. The fact is enjoyed and so conserved. This relationship can be called the **line of conservation**.

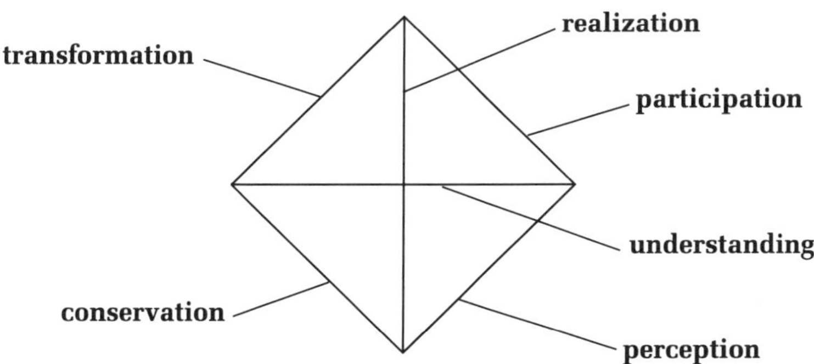

Figure 4.5   The Six Connections of the Tetrad

EXAMPLES OF TETRADS

To help readers better understand the tetrad, I want to illustrate its use in examining three different activities: gardening, musical performance, and family life. A fourth activity— governance—is presented in greater detail with consideration of the six connections among its four sources.[1] Readers should also practice using the tetrad to examine activities with which they are personally familiar and to compare their results with those below.

Gardening

The tetrad of gardening is illustrated in figure 4.6 and involves the four sources of *land to be worked, ideal garden, cultural context,* and *gardening skills.* The garden arises from a particular plot of land that through its size, topography, soil quality, and so forth, sets the ground for what the garden will become. At the same time, an image of the ideal garden gathers together the potential of the plot and directs its content and design.

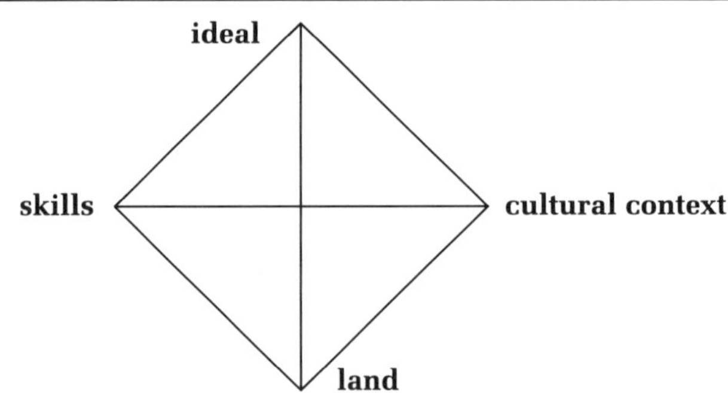

Figure 4.6   Gardening as a Tetrad

In dialogue with the image of the ideal garden is the cultural context that guides the gardener's knowledge of how gardening is done. For example, eighteenth-century gardens generally followed classical tenets and were regular, symmetrical and carefully manicured. In contrast, nineteenth-century gardens were influenced by the Romantic movement and emphasized irregularity, asymmetry and a sense of wild nature. Finally, gardening is also dependent on the skills of the gardener, who may be an inexperienced beginner or a gifted expert with the proverbial "green thumb." Gardening involves a relatively simple tetrad, and through it, one can better consider the variety of gardening activity and the importance of balancing the four sources so that a particular garden will be successful.

Musical Performance

The tetrad of musical performance is illustrated in figure 4.7 and involves the four sources of *actual sound, unheard music, interpretation,* and *execution.* The ground of performance is the actual music, which the performer creates, listens to, and judges, simultaneously. The basis for the music is both the musician's musical understanding (interpretation) and his or her training and skill (execution). At times, he or she may feel that the music is only sounds produced by physical action on the instrument. At other times, he or she feels a sense of perfection and the music "speaks." In these special moments, the sound seems to come from beyond the sensual, and the instrument is but an occasion for music to enter manifestation.

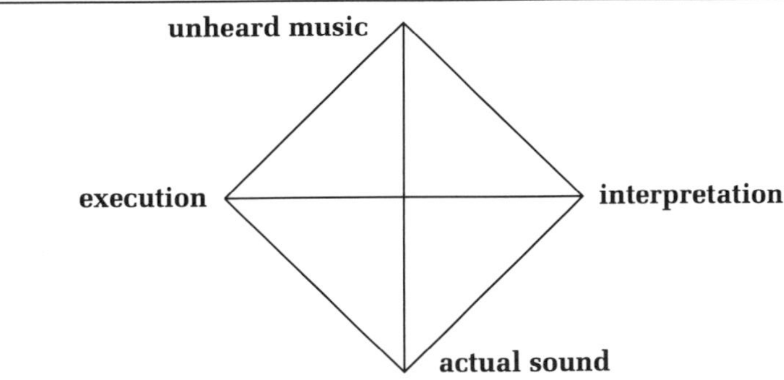

Figure 4.7   Musical Performance as a Tetrad

Family Life

The tetrad of family life is illustrated in figure 4.8 and involves the four sources of *family life, family pattern, fathering,* and *mothering.* Family life is a complex of activities that includes eating, sleeping, work, play, children growing up, parents gaining responsibility, and so forth. The family creates a home that involves a particular set of activities. In addition, the family is involved in activities outside the home. The sum of these experiences mark the ground of the tetrad.

The motivation and guide for these activities is the family's image of the ideal family, which marks the upper term of the tetrad. The means for sustaining and strengthening the family involves both father and mother. Traditionally, the mother is involved in the day-to-day ordering of the family, while the father, as "head of the household," gives the family a sense of direction and purpose.

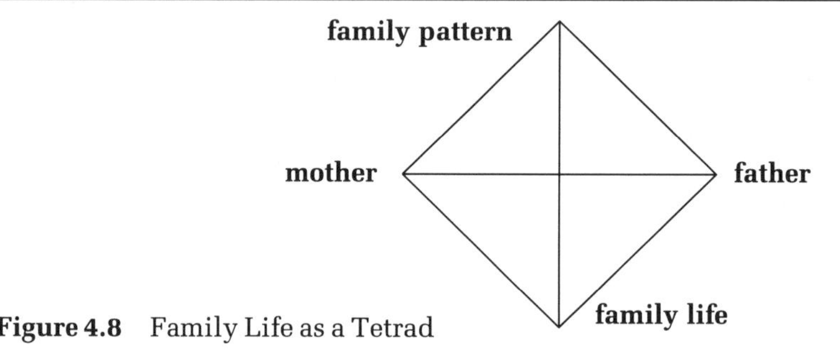

Figure 4.8   Family Life as a Tetrad

Governance

The tetrad of governance is illustrated in figure 4.9 and involves the four sources of *people, counselors, policy-makers*, and *administrators*. Whether totalitarian or democratic, a government must be able to regulate the large-scale activities of the state and direct resources toward a particular end. A government must attend to the needs of the people as well as organize and direct public affairs. How these aims are carried out is dependent on some overall conception held in place by some group of impartial "counselors", who generally do not occupy a recognized position, but are called on for advice.

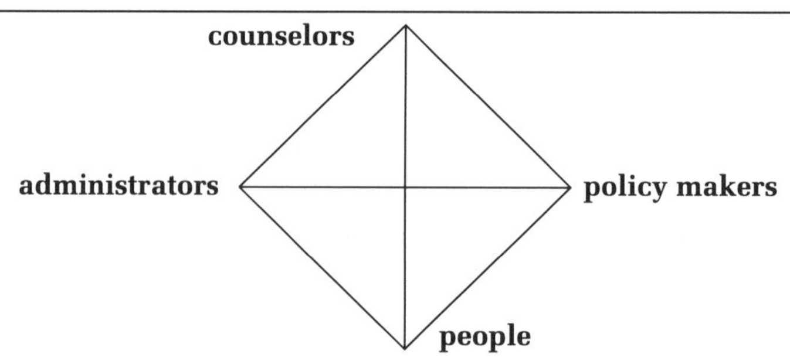

Figure 4.9   Political Governance as a Tetrad

Traditionally, governments have separated out the roles of organization on the one hand, and policy on the other. Organization is associated with the administrators, who must make sure that the governmental apparatus works practically. Policy makers determine what the administrators do and provide governmental direction.

Figure 4.10 illustrates the six connections among the four sources of governance. The connections are as follows:

1. The connection between administrators and people can be called *competence*, which involves the satisfaction of the people's needs and the organization of public affairs.

2. The connection between policy-makers and people can be called *realism*, which involves programs and reforms in touch with what the people need and can bear.

3. The connection between administrators and counselors can be called *consciousness*, which involves the qualities of impartial men and women entering the work of government.

4. The connection between policy makers and counselors can be called *vision*, which aims for the common good beyond the solution of immediate problems.

5. The connection between policy-makers and administrators can be called *cooperation*, which allows for plans that are conceived centrally but actualized in a decentralized way.

6. The connection between people and counselors can be called *social reality*, which reflects a process of maturation away from personal wants toward a comprehensive understanding of the social good.

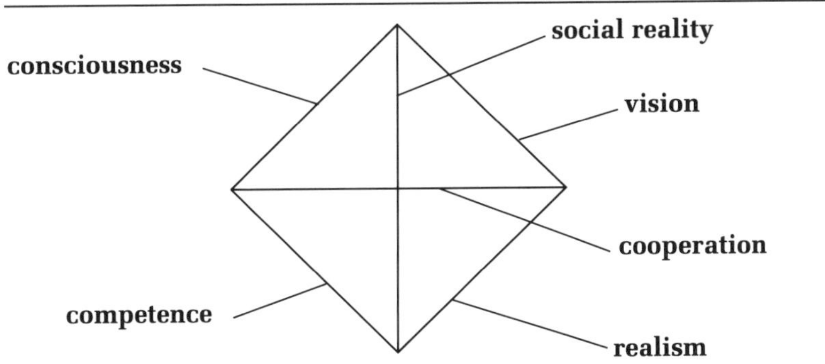

Figure 4.10   Six Connections among the Four Sources of Governance

FINAL COMMENTS ON THE TETRAD

For a complete understanding of an activity, both the simplicity of the tetrad and the complexity of the particular situation must be held in mind together. In this way, the terms of the tetrad and their six connections become more than bare abstractions or convenient fictions. One must ask how particular activities with which he or she is familiar can be expressed as a tetrad, and how the connections among the four sources can be understood.

The tetrad involves a more simple pattern than the triad, yet this simple pattern is the structure underlying all activity. Too often, the details of a situation confuse us with their complexity, yet the underlying organization of the situation may be much more simple. The tetrad provides us with a means for circumventing the confusion of details and holding sight of the larger pattern.

As was explained in the introduction, details and classification relate to *knowledge*, while structure relates to *understanding*. Like the other systems, the tetrad is an instrument of understanding, rather than knowledge, and is best used when we apply it to activities that we have thoroughly experienced firsthand. We can then relate this understanding to activities with which we are less familiar and see similar underlying patterns at work. In this way, we can understand activities about which we know little.

NOTE

1. These examples are not part of Bennett's lecture on the tetrad but, rather, are model answers given for questions in the work sheet accompanying the lecture. The editor has incorporated these answers here.

CHAPTER 5

The Pentad

WE COME TO the last system described in this book—the *pentad*, or five-term system. The pentad represents the significance of structures. It illustrates the importance of things, people, or situations in regard to their world. The first four systems we have considered—monad, dyad, triad, and tetrad—help us to understand the structure itself and how it works. The pentad is the first structure that helps us to understand why a structure works and how it has relationship to the world of which it is a part.[1]

We are never interested in something only because of what it presently is. We are also interested in its potential. When we say that something is significant or has possibilities, we do not simply mean that it is interesting in itself. Rather, we infer that the thing stands out from its surroundings because it is useful, unusual, pleasant, unpleasant—in short, because we implicitly realize that the thing has an effect beyond itself. It is a thing-in-itself, but it also has reach beyond its own limits. In other words, the thing is important for what it can do in the world—for the ways it can touch its surroundings. This ability of the thing to be more than itself leads to the words **significance** and **potentiality** as key characteristics of the pentad. These two terms have slightly different meanings, and we must find a way to integrate both into our understanding of the pentad.

ESSENCE AND CONTENT

To understand how five terms can describe significance and potentiality, we must start with the "selfness" of the thing. This thing, or structure, has a particular character and place. In this sense, the structure is a center of its own meaning and a point of significance. In other words, one way to be significant is to be oneself. When something is what it is, its significance is concentrated at a point. This centeredness of the thing generally leads to the thing having a name. A name is much more powerful than a mere word because it identifies a thing with recognized potential. In one sense, we can say that everything that exists is significant just by the fact that it is what it is. When we say, "This is what it is," we are really saying that there is a point in which the significance of the thing is concentrated. We call this point **essence**; it represents the first term of the pentad.

The terms of the pentad are called **limits**, and we can use the center of a circle to represent essence as the first limit. As figure 5.1 illustrates, we can say that the circle represents a structure and all its content, while the center of this circle represents the essential identity, or "is-ness," of the structure. For example, if the circle represents me as an individual, then the space it encloses represents my body, my feelings, my thoughts, my memories—in short, everything that I know and experience. The circle's content could be called *me*. But somewhere in this circle is a center of significance identified by the phrase, "This is I, John Bennett." This core of me as an individual is represented by the center of the circle.

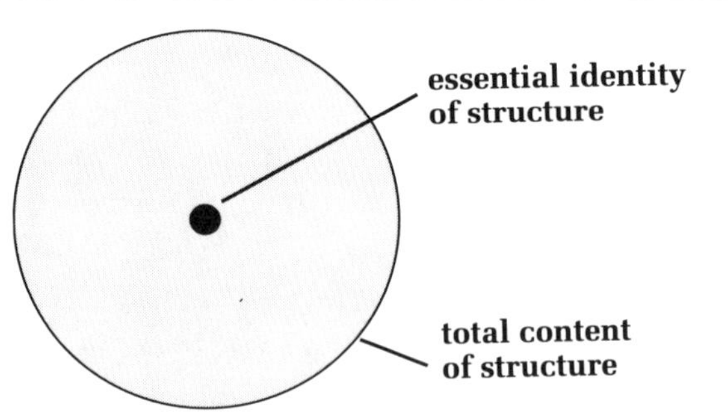

Figure 5.1   The Structure as a Circle

My name summarizes all that I am—that part of myself I know or don't know, that part visible or invisible to others, that part already existing or someday to develop. All of these parts are within the circle having the center to which my name refers. These parts could not exist if I had no central core that is uniquely me, the essential John Bennett. This central core—my essence—is the first limit of the pentad.

As another example of essence and content, consider an ordinary object like a blackboard. It, too, has a certain content that we could put inside a circle and say, "This is the blackboard and what it is composed of." We would speak of the raw materials of which it is made, its size, its style, its method of manufacture, and so forth. But there is also something that leads me to use the word "blackboard" rather than "sofa" or "chair" because all the characteristics of a blackboard—the part that it plays in human life—are somehow concentrated in this word. As figure 5.2 illustrates, we can again represent this meaning of the blackboard by a circle and its center. The complete circle represents the total content of "blackboard," while the circle's center represents those qualities that make the blackboard essentially what it is.

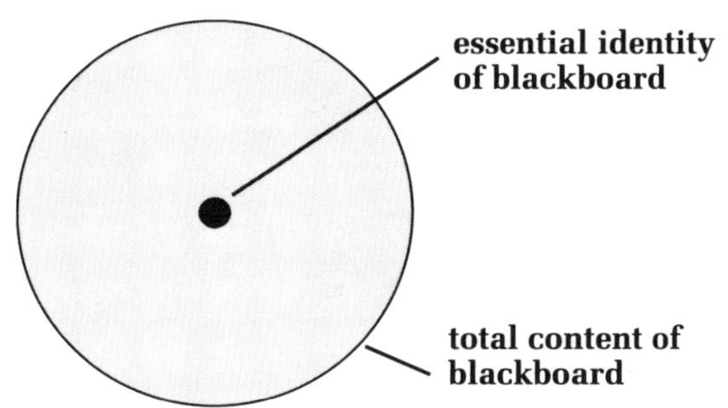

Figure 5.2   The Blackboard as a Circle

THE STRUCTURE AND ITS WORLD

But besides the blackboard's essence and total content, there is also the blackboard's relationship to its surroundings—the ways it belongs in the world. My world is the entire range of people, things, and events to which I am connected, either in the past, present, or future. Everything with which I cannot have contact is not only outside me but also *outside my world*. In this sense, something that has a small world is less significant than something that has a large world, since the latter has a wider sphere of contact, at least potentially.

The blackboard only has a world when it is being used. This world is confined to people who need to communicate. Further, the blackboard's world is limited to the building in which it is located or to other buildings to which it is moved. These limitations indicate that the blackboard's world is quite small. Yet the world of a piece of chalk is even smaller, since the chalk's power to make itself felt amounts to no more than making marks on a blackboard. Like the blackboard, the chalk has no world unless it is being used.

As human beings, we participate in a much larger world, though, clearly, some people's worlds are larger than others'. Whatever the particular structure whose world we consider, we can represent its relationship with the world by a second, larger circle surrounding the smaller circle of figures 5.1 and 5.2. The result is illustrated in figure 5.3. Note that the region beyond the second circle is outside the world of the structure being represented.

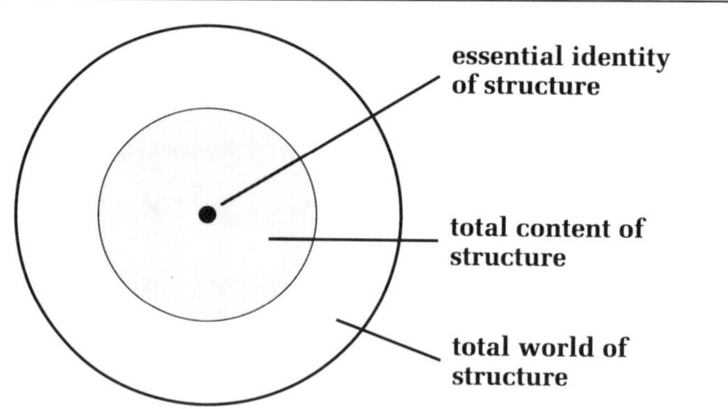

Figure 5.3   A Structure and Its World

SOURCE AND END

But how do the concentric circles of figure 5.3 relate to the pentad and five-ness? If anything, the figure suggests three-ness—the two circles and their common center. If essence as the common center represents the first limit of the pentad, where are the other four? To answer this question, we must realize that the world in which we find ourselves has many different influences on our lives. In one way, we arise *from* our world; in another way, we complete ourselves *in* this world. In terms of significance, therefore, we can speak of the world in two ways: First, the world as **source** from which we come; second, the world as **end** toward which we aim. These aspects of the world represent two more limits of the pentad.

As a source, the world is, first of all, that from which I arise. My heredity, constitution, geographical environment, cultural tradition, and so forth—all these qualities come from one small part of the greater world and help make me who I am. At the same time, however, I seek fulfillment. This fulfillment cannot embrace the whole world, yet it points toward my end, not in the sense of some physical place but in the sense of completion, as in the Greek word *telos*, which means the destination toward which a life goes.

Figure 5.4 illustrates these two distinct ways in which I derive personal significance from the world in which I belong. The lower point on the outer circle represents the *source* from which I come, while the upper point on the outer circle represents my potential *end*. We now have identified three limits of the pentad.

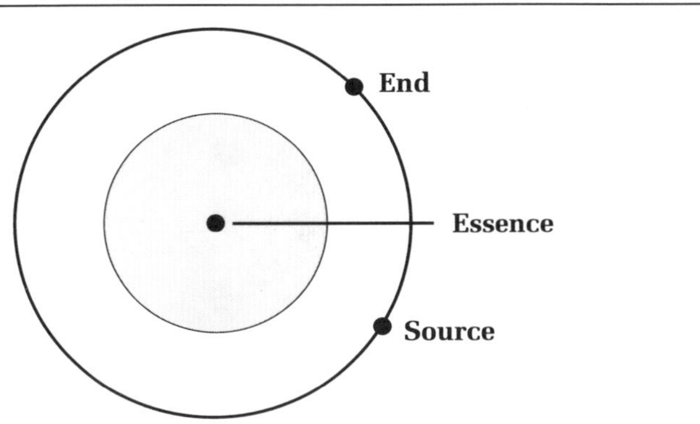

Figure 5.4   Essence, Source, and End

FACT AND VALUE

To locate the last two sources of the pentad, we must consider in more detail our relationship with the inner circle, which represents the total content of a thing. One way in which I can understand my own significance is by the simple fact that I exist as a living body. If I am to guarantee myself bare sustenance, I must make sure my material body can survive. This minimal basis for my continued existence can be called my *factual nature*, or **fact**. A person's factual nature is visible and includes a physical body with certain functions and powers.

At the opposite extreme, I also believe that there is some inner ideal toward which I can grow as an individual. I have certain inner possibilities that point to potential self-perfection. This image of my ideal self can be called my *potential nature*, or **value**, which is considerably different from my factual nature in the world. A person's value is largely unseen and points to the hidden side of his or her nature that draws the person upward toward greater aims.

We have, therefore, two extremes within the totality of my selfness. These two extremes are the limits of the inner circle and can be represented graphically as in figure 5.5. First, there is the structure as minimal entity, which is represented by the inner circle's bottom point and refers to *fact*. Second, there is the structure as it has possibilities, a quality represented by the inner circle's upper point and called the structure's *value*.

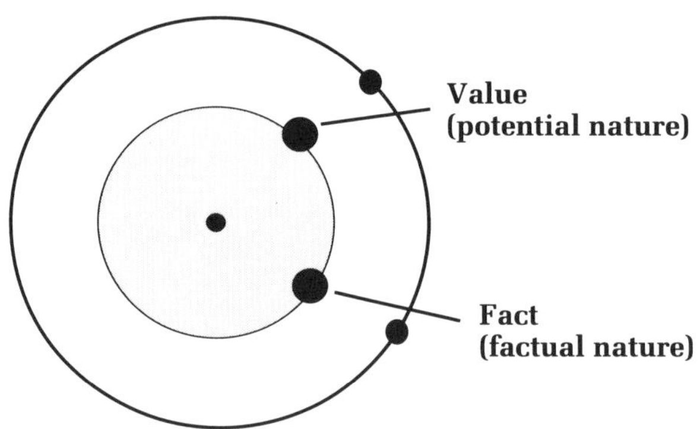

Figure 5.5   Significance, Fact, and Value

All structures have value. The blackboard, for example, is not only a piece of equipment with a factual nature. The blackboard has considerable value that lies in its ability to facilitate communication. But note that this potential is not necessarily equivalent to its end, which relates to what the blackboard will have actually accomplished in the world of communication by the time it is discarded. This end relates to the benefits the blackboard has provided for the teachers and students who are destined to use it. Yet if the blackboard breaks or is stored away or is without a supply of chalk, its value will not approach its end. A structure's value points to possibilities that may or may not be actualized in regard to its larger world.

RECIPROCITY AND THE PENTAD

Figure 5.6 summarizes the five limits of the pentad. As the centering limit of the pentad, the *essence* relates to the very heart of the structure, while *fact* and *value* reflect the range of the structure as a thing-in-itself. As the two outer limits, *source* and *end* reflect the structure's range of being in relation to its larger world.[2]

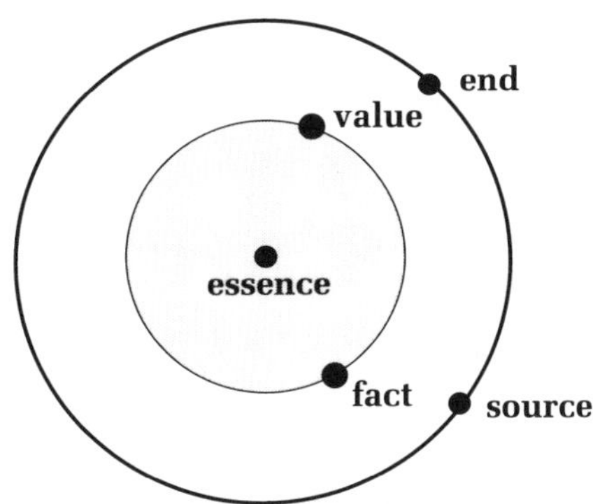

Figure 5.6 The Five Limits of the Pentad

Figure 5.6, however, represents the five limits as isolated points and, therefore, suggests no links among them. In fact, the links among the limits are crucial, and they can be represented by the symbol of figure 5.7, which is the completed pentad.[3] It appears as an enclosed, five-pointed star comprised of ten lines connecting the five limits. These ten lines are important because they represent the *reciprocity* of the pentad.

This reciprocity is another key to the significance and potentiality of a structure. An examination of the ten connections among the five limits provides an understanding of this reciprocity. Before we look at these connections in detail, however, I want to provide a general picture of reciprocity by considering a children's school as a pentad.

                    Elementary Systematics

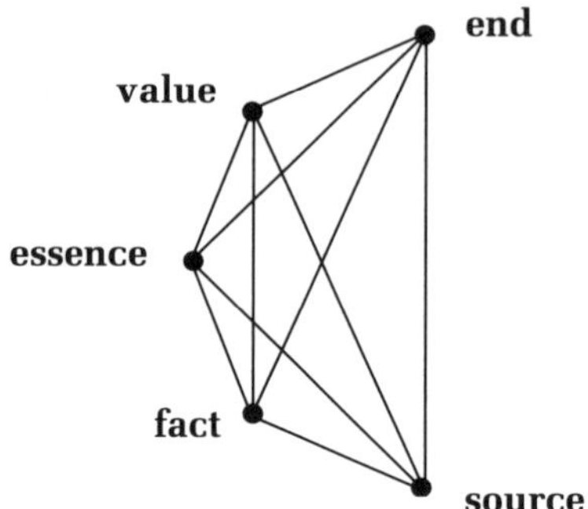

Figure 5.7   The Pentad and Its Five Terms

Figure 5.8 lays out this school as a pentad and identifies the limits of this school. Its essence is its "educational atmosphere"—the tenor of feeling and set of attitudes that influence relationships and mark the heart of the school's educational aims. Fact for the school is "minimal learning skills," while value involves a "balanced development" of thinking, feeling, and action. The source of the school is an "uncultured society" and its uneducated children, while the end is a "cultured society" composed of mature adults.

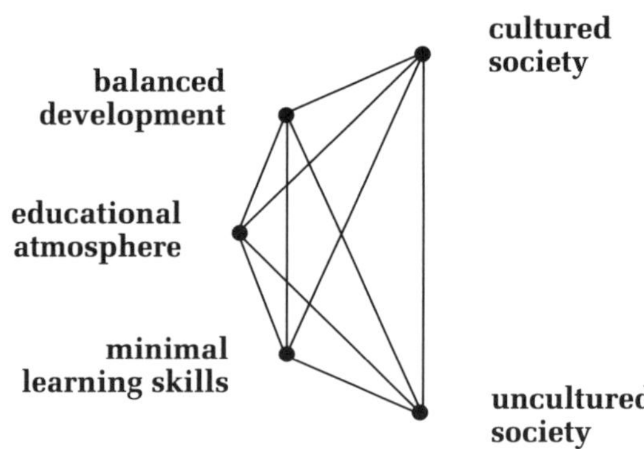

Figure 5.8   Pentad of a School

One readily recognizes that these five limits of the school cannot be isolated from each other but must involve exchange and reciprocity. In general terms, this reciprocity can be described as the upward and downward flows shown in figure 5.9. The upward flow is *transformation*, which the school accomplishes by raising the cultural level of the children, either through general education or some specialized training. For this upward flow, however, there must be a reciprocal downward flow, which comes about as the school looks toward its end—a cultured society—and sets educational aims that are challenging yet realistic. These aims *inspire* the uncultured society, which remains what it is but contributes individual students for the work of the school. The transformation of individuals represents the upward flow of the pentad, while the inspiration given by the school to its source represents the downward flow.

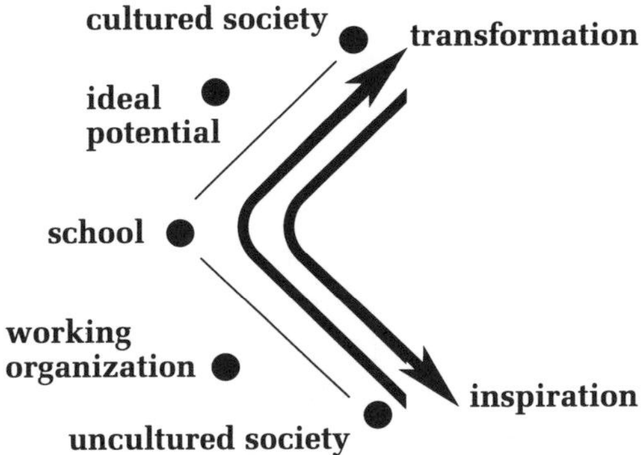

Figure 5.9   The Reciprocal Flows of the Pentad

The reciprocity among limits in these upward and downward flows helps one understand why Gurdjieff associated the pentad with what he called "reciprocal maintenance" or "reciprocal feeding." [4] If, for example, we consider the connections among the school's essence, source, and end, we realize that the school as "atmosphere" assimilates the "uncultured society" and, at least figuratively, "feeds" on it. In this sense, the essence "eats" the source. From another vantage point, however, the school's end is fulfilled through the essence; here, essence is "food" for the end.

INNER AND OUTER SIGNIFICANCE

Another way to illustrate reciprocity in the pentad is to speak of a structure's inner and outer significance. Everything has an interior life and an exterior life, a life for itself and a life for others. Both inner and outer lives have a significance and potential that can never be divorced without losing the essence of the thing. On the one hand, a thing valued just for what it is and not for the purpose it serves, loses both its place in the world and much of its meaning. On the other hand, if the thing is considered only for its utility and not for what it is, its integrity as a thing falls into question. It is not respected for what it is, and its value is diminished or destroyed.

The importance of a mutual relationship between inner and outer significance can be seen in the connection between *essence and end*. If the relationship were only one way from essence to end, then completion would be annihilation; essence would be swallowed by end and inner would be overwhelmed by outer. Some spiritual traditions see human destiny this way: each individual disappears at death, no longer existing as a separate entity but part of some larger whole. This vision of destiny minimizes the significance of the individual, who is subsumed by a greater totality.

Such a uni-directional relationship, however, ignores the fact that the pentad's limits are always reciprocal. In regard to the link between essence and end, this mutuality means that the more fully a thing is itself, the more fully it can accomplish its end. Conversely, the more fully a thing accomplishes its end, the more fully it becomes itself. The reciprocity between essence and end is integral to all structures and must always be kept in mind if we seek thorough understanding.

One also sees this reciprocity in the connection between *essence and source*, which is not a thing that we depart from never to return. If we sever ourselves from the source, we lose connection and forfeit participation in the reciprocal flow between source and end. We must always have concern for the source from which we arise; this requirement relates to the condition of our having a significance *above the level of the source from which we came*. Always, we should pay the source for our indebtedness by returning something to it.

THE CONNECTION BETWEEN FACT AND VALUE

The connections that essence makes with source and end help us to understand the relationship between a structure's inner and outer significances. Next, I want to consider the connection between *fact and value*, which helps us to understand connections among aspects of the structure itself. From one point of view, human beings can be observed and studied from the outside as other living things of nature can. This external picture presents the person's *factual nature*. Some schools of thought, like traditional behaviorist psychology, study human life only in observable terms as a complex system of facts and material relationships. This visible picture of human nature, however, is incomplete, since human life also involves an unobservable inner dimension. A picture of human beings that involves only their outer expression ignores many dimensions of human life that cannot be studied materially, yet are crucial for a complete understanding—for example, motivations, expectations, feelings, and so forth. These less visible aspects of human life refer to *value* in the pentad and indicate that people are more than just fact.

We can represent the reciprocity between a person's outer and inner dimensions by figure 5.10, which incorporates what the person inescapably is (fact) as well as what he or she might become (value). One must remember that fact and value also apply to other structures besides human beings. Value exists for anything that has significance or interest, including material objects. If a thing has significance, it is more than a fact, since significance cannot be reduced to factual terms. All facts are equal, which means that the quality that makes one thing more important than another belongs to some realm other than facts.

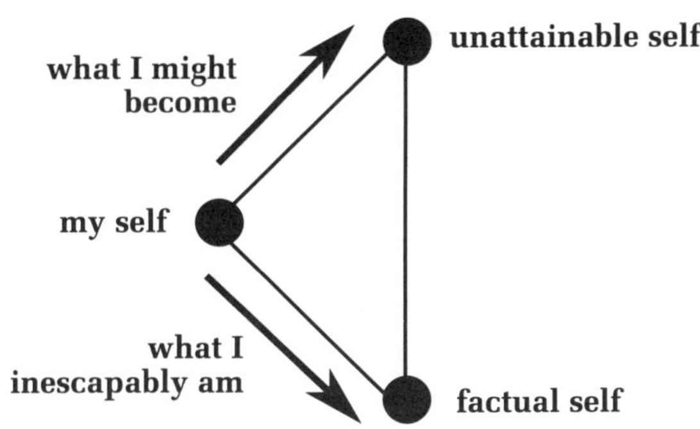

Figure 5.10   Reciprocity betwen Factual and Unattainable Selves

This realm involves value. For the human being, value refers to the kind of being that the person is given the possibility of becoming. This possibility is more than just what one is in the present moment. Rather, value points to something unattainable—a certain perfection making demands that by their very nature can never be fully satisfied. In figure 5.10, this limit is called *unattainable self.*

To understand the significance of a structure, we must take into account its value. What might the structure become? What are its hidden possibilities? Sometimes these possibilities are obvious because the thing has gone a long way toward realizing its potential. This actualization of possibilities, however, is not the important point in terms of the pentad. Rather, one seeks to clarify the structure's potential, whether actually carried through or not.

Another way to understand the relationship between fact and value is to say that the structure derives its strength, on the one hand, from the inescapable fact of its own existence and, on the other hand, from the unattainable demand of its potentialities. Again, let me illustrate with the example of the human being. To be successful, I first must know myself as a fact—I must be aware of my physical assets and limitations. At the same time, however, there is an inner pattern—one might call it an ideal or dream—that never leaves me satisfied. I sense my distance from this "other" that is not outside and different from me, but a part of me, though ideal. I cannot call this "other" the center of myself, since it is unrealized and unattainable.

The vertical line between fact and value represents all levels of existence that are present in a structure. These levels of existence range from elements of quantity and function to elements of quality, wholeness, and depth. In the case of the school, we can establish its practical function through statistics and other information concerning attendance, grades, number of scholarships, and so forth. In this sense, we describe the factual end of the fact-value connection. As we ascend this line, we move into the value aspects of the school—its deeper significance and potential. In what ways, for example, have graduates of the school been successful in life? What kinds of feelings does the school evoke for students, teachers, alumni, and other people associated with it? Overall, this line represents the range of educational processes from general outer instruction to individual inner integration.

Figure 5.10 illustrates the mutual actions among essence, fact, and value as applied to the pentad of the individual human being. The reciprocity among "self," "factual self," and "unattainable self" always underlies human experience. On one side, there are the facts of our existence. On the other side, there is the call of our potentialities. And, at the center of it all, is our essential self that can remain the same, fall downward, or strive toward possibilities.

CONNECTIONS BETWEEN FACT AND SOURCE AND VALUE AND END

There are several other connections to examine within the pentad, and we start with the links between *fact and source* and *value and end*, illustrated in figure 5.11. We can clarify the relationship between *fact and source* by returning to the example of the blackboard. Its source includes the stone and wood that establish its factual nature. In a physical sense, the potential of these source materials is now realized, since the blackboard can now only be what it is. Through the fact, the possibilities of the source become actualized.

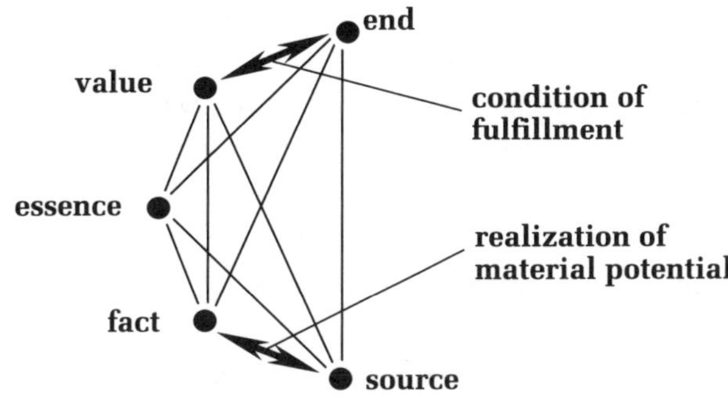

Figure 5.11   Connections between Fact-Source and Value-End

Figure 5.11 also illustrates the connection between *value and end*. This link means that a structure's value is intimately related to the end in which it finds its fulfillment. As I explained above, the value of the blackboard is its effectiveness as a medium of communication; this value is worthless if the blackboard is stored away and never used. The blackboard finds its end only in the world of human communication. In this sense, a structure's value is intimately concerned with its fulfillment in the world. Note, however, that this connection between value and end involves a *condition* of fulfillment, while the connection between essence and end involves the *reality* of the fulfillment.

THE CONNECTION BETWEEN SOURCE AND END

The connection between *source and end* identifies the range of possibilities in regard to the world in which the structure finds itself. A structure cannot participate in either the realm below its own source or in the realm beyond its own end. The significance of anything, so far as it is what it is, has a terminus marked by the connection between source and end. The significance of the blackboard, for example, cannot extend beyond the world of human communication, since outside that world the blackboard is no longer itself. In the same way, the blackboard's significance cannot reach beyond the world of raw materials. To discuss the blackboard in terms of wave mechanics or theories about the ultimate nature of matter is pointless, since these themes have no bearing on the significance of the blackboard as a blackboard.

In regard to humanity, the limits marked by source and end mean that human destiny has a certain terminus. This terminus, however, is not a final stopping point but a place where humanity is transformed to participate in further potentialities, though not as human beings but as something else. To ponder essence, standing before the line of source and end, is to contemplate the human being's place in the world and cosmos. Beneath essence is the entire birth of life from which humanity arises; above is the consolation of all human destiny.

The connection between source and end is one of the two vertical lines of the pentad. The other vertical connection is that between fact and value, which we have already discussed above. Figure 5.12 summarizes these two vertical connections, which together represent the range of a structure's possibilities in terms of its inner and outer aspects. On one hand, the line between fact and value speaks to the structure's own levels of existence. In this sense, one can say that the smaller triangle marked by essence, fact, and value represents what the particular structure is and what it might become. On the other hand, the line between source and end speaks to the structure's range of potentialities in the world. In this sense, the larger triangle marked by essence, source, and end represents the world to which the structure belongs or can belong, either actually or potentially.

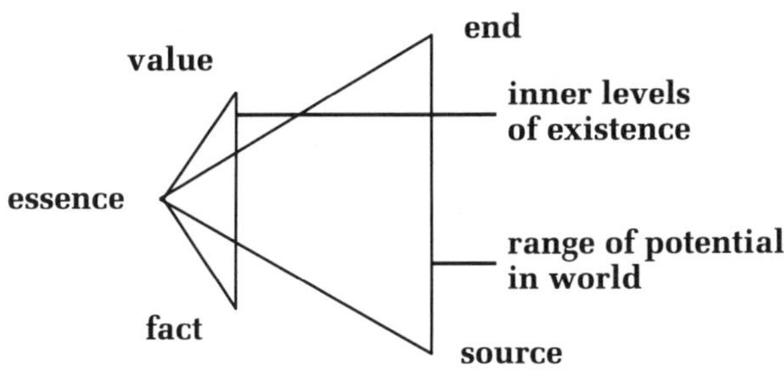

Figure 5.12 The Pentad's Vertical Connections

CONNECTIONS BETWEEN SOURCE AND VALUE AND FACT AND END

The connections between *source and value* and *fact and end* are illustrated in figure 5.13. The connection between *source and value* gives meaning to the source as it can be a potential medium for values. This meaning is transcendent to the source's fulfillment in the essence, since the value here is beyond the limits of the source's own world. This connection demonstrates that the full significance of a structure includes a power by reason of its highest potentialities. Through this power, the source is given a transcendent meaning that is structured out of fact and assimilated into essence.

In contrast, the connection between *fact and end* represents the necessity of fact in the attainment of fulfillment. Without the fact, there can be no concrete realization of destiny; it is through the fact that the dream of attainment becomes actual. The connection between fact and end means that human beings' inescapable "is-ness" is essential for the realization of human destiny. At the same time, the connection between source and value says that the world out of which human beings come can have a transcendent meaning, since that source is the medium from which all human potentialities originally arise.

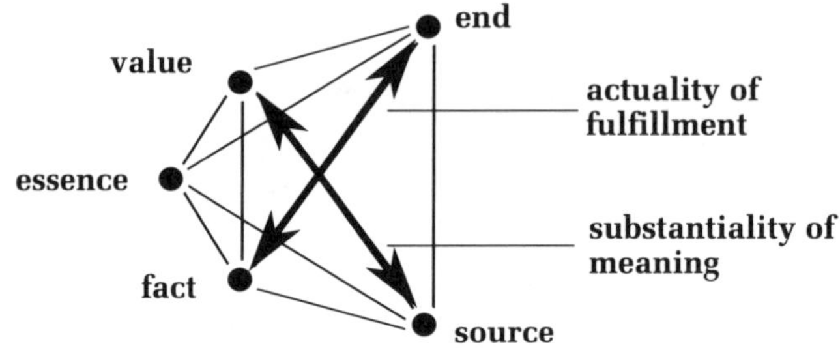

Figure 5.13   Connections between Source-Value and Fact-End

USING THE PENTAD

The pentad is the first system in which the structure is an *independent totality*. Unlike the monad, dyad, triad, and tetrad, the pentad does not ignore the rest of the world in considering the particular structure. Rather, the independence of the structure is found together with its dependence on the world to which it belongs. The pentad of the blackboard, for example, draws together its intrinsic significance and its significance for the worlds of raw material and human communication.

In working with the pentad, I have found that it can produce a change in one's inner state that is really most extraordinary and quite unmistakable. The pentad allows our attention to hold a deep sense of a structure. We encounter a fullness of reality that we ordinarily miss. This way of seeing, however, is not easy and requires much effort and practice.

One of the best ways to work with the pentad is to use it to consider a thing that one finds personally important in his or her life. First, consider the thing factually as it can be seen from the outside. Next, consider the value of the thing: To what perfection can it aspire? What are its greatest possibilities? Third, take into account the essence of the thing by asking how much power it has to hold its factual and value natures together. Last, relate the thing to its source and end by asking how it arose and to what farthest destiny it might go.

If one considers a structure in this way, he or she finds that little by little its meaning begins to stand out in relation to its world. There appear qualities of significance that one had not suspected before. At the same time, one is protected from false imagination about the thing.[5]

A FINAL COMMENT

Whether we study a structure in terms of monad, dyad, triad, tetrad, or pentad, we can gain a full understanding only when we have understood from *our own experience*. There is nothing wrong in borrowing from other peoples' interpretations or knowledge, since in these sources you can often recognize patterns with which you are familiar. But I must emphasize that, in the end, *systematics cannot take the place of experience*. Fortunately, we all have more experience than we have understanding, and, therefore, for all of us there is some resonance that we can transform for any subject toward which we turn our attention. In short, systematics enables us to derive maximum understanding from personal experience—to discover clear, coherent patterns in the midst of life's complexity and ambiguity.

NOTES

1. In volume 2 of *The Dramatic Universe* (p. 12), Bennett clarifies these groupings of structures in greater detail: "It has been found that for purposes of practical utility, the systems fall naturally in groups of four. The first four from the monad to the tetrad help us to see *how* structures work. The systems from pentad to octad show *why* they work and how they enter into the pattern of Reality. The third group from the ennead [nine-ness] to the dodecad [twelve-ness] is mainly concerned with the *harmony* of the structures: that is, the conditions that enable them to fulfill their destined purpose." He continues: "The pentad can be understood from its place at the start of a new cycle within the series of systems. The first four systems, with the attributes of universality, complementarity, dynamism and activity, are all abstract in character. They do not particularize the structure to which they apply. With the pentad, we have a new property given by the character of the terms as limits: the possession of a boundary separating "within" and "without." The structure thereby becomes an *entity* that can be known and understood by its inner nature and its outer connections. From this observation, it seems to follow that we cannot regard entities as simple notions definable by a single characteristic, but rather as structures having at least the complexity of the pentad" (p. 42).

2. In his later work, Bennett provided different names for the five limits: essence became **ipseity**; fact, **lower nature**; value, **higher nature**; source, **nourishment**; and end, **master**. See *The Dramatic Universe*, vol. 3, pp. 40-41.

3. In earlier work, Bennett used a four-pointed star as the symbol of the pentad; the center of the star represented essence (see *The Dramatic Universe*, vol. 2, pp. 296-297). He attributes the present symbol to the Elizabethan physician and alchemist Robert Fludd and his *Utriusque Cosmi Maioris et Minoris Metaphysica* (ibid., vol. 3, p. 41).

4. G. I. Gurdjieff, *All and Everything: Beelzebub's Tales to His Grandson* (New York: Harcourt Brace, 1950); also, see J. G. Bennett, *Gurdjieff: Making a New World* (New York: Harper & Row, 1973; Santa Fe, New Mexico: Bennett Books, 1992), chap. 9 and appendix II.

5. Other than the school (figure 5.8), this lecture provides no illustrations of specific pentads. In the answers to the exercises following lecture 5, however, several pentads are discussed, though in a skeletal way. Realizing that readers might find these descriptions helpful, the editor includes three here: the pentads for "factory," "library," and "school."

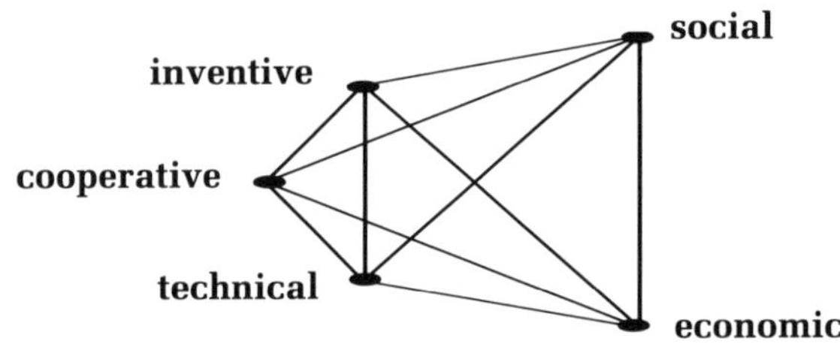

Figure 5.14 Pentad of a Factory

Explication: The world of a factory involves physical operations and their technical basis, interactions among workers and managers, and research and development. Loyalty, efficiency, and inventiveness are key factors for the successful factory. For society, the factory is significant both economically and socially. The standardization of production and employment of many people is crucial for the continuing success of modern societies. The factory has social obligations to society as a whole.

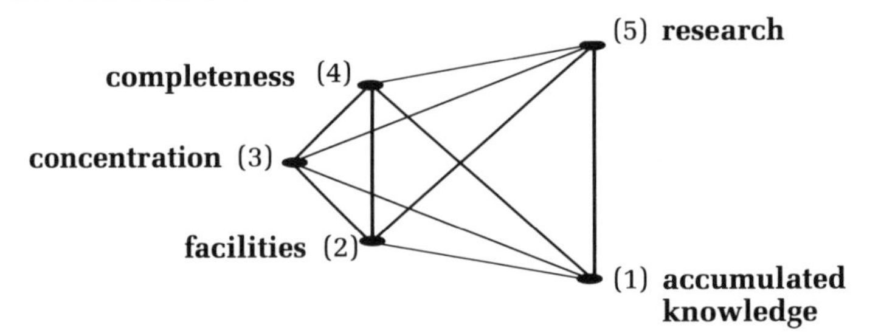

Figure 5.15 Pentad of a Library

Explication: We assume a research library that plays a significant role in the world of acquiring and enhancing knowledge. The library's source is the accumulated knowledge of humankind. This source provides material that the library catalogues and preserves. Each research library has a unique concentration of materials that makes it a specific source for the world of research. Because of its particular specialization, the library assumes ideals of completeness that are independent of the needs of research. The library should hold or have access to a truly representative collection of materials relating to its specialization.

The facilities of the library determine the availability of the material (2-5), access (2-1), and capacity (2-3). On one hand, the library should remember its ideal of completeness and maintain its unique character. On the other hand, the library that does not render a service to research becomes moribund and museum-like.

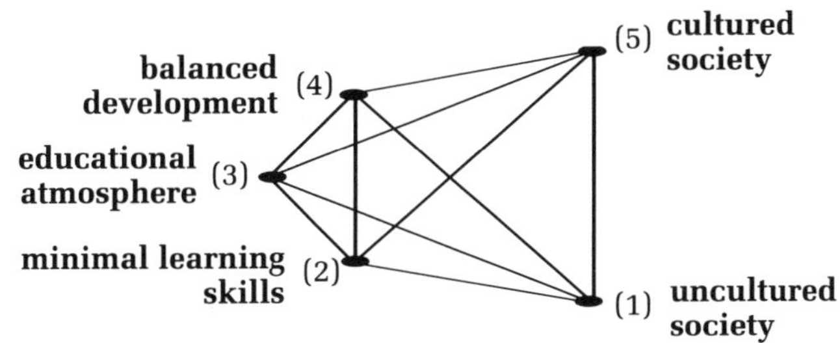

Figure 5.16 Pentad of a School

Explication of the Ten Connections for the Pentad of the School:

1—2: Admission and development of personality;

1—3: Participating in school society and meeting personal challenges;

1—4: Integration behind the outer instrument of personality; the beginning of deeper personal development;

1—5: Range of action for human being in the world;

2—3: Care over study and work;

2—4: Range of educational experiences from general outer instruction to individual inner integration;

2—5: Foundation for carrying out duties in life;

3—4: Flexibility and companionship; encouraging the individual's spontaneity and giving attention to a diversity of needs;

3—5: Respect for human values above scholastic achievement;

4—5: Foundation for a fuller self.

General Systematics

THIS PAPER ARGUES that understanding is possible because there is a world order that is reproduced or reflected in our experience through systems or sets of terms having universal properties. *Systematics*, or the study of systems, should, according to this thesis, be the appropriate instrument for the development of understanding, as science is the instrument for the development of knowledge.[1]

Understanding is a special relationship between different parts of our experience. There is the general presentation of our senses that we sometimes call the "world." There is also an inner awareness that is called thought, or at least includes thought. Apart from these two ways of coming to meaning, there is also an experience of directing, selecting, and choosing between the various presentations that we associate with attention or the power of attention. Finally, there is the experience of willing—that is, initiating action through our own bodies to change either self or world. Some combination of these four modes of experiencing gives rise to what we call understanding.

Understanding differs from sensation and thought by the property of standing astride, as it were, the various parts of the total experience and knitting them together. When we understand, our knowledge is linked to our sensation; but a connection is also made between our inner attention and outward actions. We cannot simplify this situation by omitting any of the four elements of sensation, thought, attention, and action. It follows, therefore, that understanding itself is not simple but possibly even more complex than this first analysis suggests. It seems likely, for example, that understanding is progressive, to be approached by stages, but must, in its fullness, always remain beyond our reach.

The impulse to understand, and not merely to know and to act, is an impulse characteristic of human beings and apparently not shared by other animals. I am not concerned here with the origin and nature of this impulse, but with the implication that there is something to be understood and that understanding is not reducible to knowledge and action. We know facts by way of perception and conception. We act from instinct and desire directed by knowledge. But it also seems that knowledge and action would be automatism—indistinguishable from animal behavior or even the work of a machine—if not informed by some kind of understanding.

The assumption that there is something to be understood beyond fact and feeling means that we suppose there is some universal order or principles in which both we and our world participate. If there were no such order or principles, caprice would reign. Anything and everything would be possible, and nothing could be known or foreseen. No one doubts that there are possibilities and impossibilities—that is, that there are universal laws that distinguish what can happen and what cannot happen. Most people would be prepared to admit that there are knowables and unknowables—i.e., that there are limits to what the individual can know of the totality of which our human experience forms so small a part. It would also probably be agreed that there are predictables and unpredictables. More generally, there are various regularities in our experience that cannot be accounted for solely in terms of the peculiarities and limitations of our human instruments of sensation and thought.

SYSTEMS AND SYSTEMATICS

In the present paper, I suggest that these universal regularities are not the result of obscure properties of nature that can only be discovered by centuries of research. Rather, I argue that these regularities involve modes of connectedness common both to the world and our experience of it. We perceive and understand in certain ways because we and the world are constructed in certain ways. These ways can be described in terms of **systems**; the study of all the possible forms of connectedness can therefore be called **systematics**.

The word "system" is commonly used for every kind of group or collection of interacting things or ideas. The meaning common to nearly all uses of the word is that of an inner connectedness that distinguishes what forms part of a system and makes it different from all that is "outside" it. I define **system** in a different, more precise, way as *a set of distinct but mutually relevant terms*. A **set** means *a limited, well defined number*—not just an indefinite collectivity. **Term** refers to *any part of experience that can be identified by some recurrent property*. A term may be a thing, an ideal, a relationship, or a complex of things, ideas and relationship—provided that we can recognize it in some way. **Distinct** means *distinguishable in regard to some property or quality*; in other words, no two terms of a system can be identical or even nearly so. Finally, **mutually relevant** refers to the condition that *each of the terms of the system requires all the others to be what it is and to mean what it means.*[2]

This definition of a system may seem so arbitrary and restrictive that we would be unlikely to meet with any set of terms that satisfy all of the requirements. In reality, we cannot perceive, think, feel, act, or understand anything except through the distinctness and mutual relevance of the objects and ideas that present themselves to our awareness. The properties or attributes of systems are the elements of all possible understanding.

Whenever we say "this is a ...," we point to a one-term system, or **monad**. Whenever we say "A is composed of B and C," we point to a two-term system, or **dyad**. Whenever we speak of connections or relatedness, we affirm the reality of a three-term system, or **triad**. Three terms are important here because, if A and B are connected, then there must be some third term C to provide the connection. A, B, and C must be distinct; the fact that they need one another to exemplify connectedness means that they satisfy our definition of a system.

In summary, systematics refers to *the study of systems and their application to the problem of studying ourselves and the world*. Systematics has four branches of study:

Formal systematics, which studies the properties of systems without reference to the nature of the terms. Formal systematics consists mainly of the investigation of possible modes of connectedness that evidently can be complex for systems with more than three or four terms.

Pure systematics, which identifies and describes the universal properties and attributes common to all systems. Later in this paper, I provide examples of the results obtainable through the study of pure systematics and describe the one-term through the eight-term systems.

Applied systematics, which studies the systems that comprise human experience and gives particular attention to their terms and their characteristics.

Practical systematics, which applies the understanding of systems to problems that arise in all departments of life.

There are three kinds of systems: **determinate, indeterminate** and **infinite**. The **order** of the system is given by the number of terms. We shall mainly be concerned with **determinate systems**—that is, systems in which the number of terms is known and less than twelve. **Indeterminate systems** can also be called **societies**. These systems are composed of a relatively large number of terms, and it is, therefore, impossible to specify all the connections or even the main groups of connections.[3] The **infinite system** is the complete expression of all possible modes of connectedness, without limits to the number and variety of terms.

The independence of the terms of a system is expressed by saying that every term has a **character**. The combination of characters gives the system an **attribute** that will depend upon the way in which the terms are connected and interact. For example, the two-term system "man/woman" has the characters of masculinity and femininity. These characters may be combined to produce harmony, mutual completion, and fecundity, or they may result in opposition, conflict, and mutual destruction. Nevertheless, there is for all the systems of a given order, a certain commonality in their total quality. This commonality leads to the basic postulate of pure systematics:

All the systems of a given order participate more or less in a common property called the **systemic attribute**. The systemic attributes are the source of the basic regularities that we discover in our experience and the key to understanding ourselves and the world.

THE MONAD

We shall now examine some of the simpler systems with a view to elucidating their systemic attributes. The attribute of a system depends upon its **order**—that is, upon the number of terms. Any system with only one term is a **monad**, which is characterized by wholeness without inner distinctions. A whole may be diversified, but the diversity has not developed into distinctions. Thus, the system remains monadic. The total content of a moment of awareness is an example of a one-term system. The whole universe, regarded as the totality that presents itself to our experience, is a monad. The combination of wholeness and diversity is expressed by the attribute **universality**, which is shared by all one-term systems. Moreover, universality can have no meaning except in a monad. A pair of objects or ideas, for example, can never be said to be a universe or involve universality, since there are many more of these entities in the world.

The only exception to the proposition that every whole combines unity and diversity is the hypothetical "true atom." It is of considerable philosophical importance that physical science tends to regard even the so-called ultimate particle as composite and indeed as a universe. The nearest notion to an ultimate indivisibility is Planck's "quantum of action." As Bohm demonstrated, there are grounds for supposing that sub-quantum levels exist.[4] The notion of *hyle*, which I developed in the first volume of my *Dramatic Universe*, suggests that universality characterizes the ground state of matter no less than the totality of its parts.[5]

Here, I should reply to the objection that a "diversity in unity" is a dualistic notion inapplicable to the monad. This objection results from a defect in language. In reality, unity in diversity is a simple, indivisible idea that should be conveyed by the word "universality." In his *Advancement of Learning*, Bacon uses the word to mean the unification of knowledge, the *Philosophia Prima* that seeks the "unity in diversity" of all the arts and sciences.[6]

We should, however, think of "unity in diversity" as the immediate and simplest delivery of our experience. The "here and now" of the present moment is both one and diverse, and there is no question of separating the unity and diversity into distinct concepts until we distinguish them. This requirement means that we must go from a one-term to a two-term system. The view that universality is a simple notion—indeed, the simplest possible notion because the most immediate—is confirmed by Piaget's studies of the development of intelligence in the child. Notions of homogeneity, continuity, undivided unity, and one as the unit of counting are all derivative and far more complicated than at first appears.[7]

I shall make the assumption that all properties common to all systems of a given order can be expressed as the system's **systemic attribute**. There should be one and only one attribute for each order, but to express this attribute it may be necessary to assemble many partial or incomplete descriptions. We have our first systemic attribute in the **universality** of all monads.[8]

THE DYAD

We come to the second-order system, or **dyad**, when we introduce a distinction valid for the whole monad. We now have two parts that satisfy the conditions of independence and mutuality. This distinction is exemplified in the account of creation in Genesis and the Babylonian tablets. "Light" is the undivided awareness of what is and corresponds to the moment of consciousness in our current experience. Light's only attribute is universality. The second step is separation of light and darkness, or of the upper and lower regions. That which was one is now two, but the two comprise the one and their separation is an enrichment and creative act.

It is difficult to see that the systemic attribute of the dyad is complementary rather than polar. Light and darkness are not opposites but **complementarities**. In the story of creation, light first appears alone, revealing the universe but not revealing distinction. When light and darkness are separated, all other dyads are revealed. Nothing can be known except by light and dark—that is, by contrast.

The notion of complementarity is of unlimited generality. It has in recent years played a great part in the development of physical theory as illustrated in the work of Niels Bohr, Max Born and others. Complementarity implies acceptance of the irreducible character of the dyad. We encounter many complementary dyads in philosophy, science, and human affairs. The Cartesian doctrine of two substances is derived from the dyad "mind/body." The dyad "man/woman" is the foundation of human society.

THE TRIAD

The dyad and its systemic attribute of complementarity gives us one key for understanding the particular structure in which we are interested. In spite of its importance, however, complementarity does not exhaust the attributes of systems. As we found it necessary to move from the monad to the dyad, so we must seek in a third term for the resolution of complementarity. This search leads to the third-order system of the **triad**.

The resolution of complementarity can be stated thus: if monad X is divided into two mutually exclusive P and Q, how is a link between P and Q to be found without reverting to the universality of X? The link must clearly be different from P, Q, or X itself. We must then postulate a triad ABC such that $A + B + C \mid \mid P + Q \mid \mid X$, where the symbol $\mid \mid$ does not mean equal or identical, but "referring to the same situation." For example, we can have $X =$ humankind, $P + Q =$ man and woman, $A + B + C =$ parents and child.

P and Q in becoming A and B have changed character. They are no longer a complementary dyad but two terms in a triad. P and Q are both human beings and thus both belong to the one-term system of "humankind $= X$."

Another example can be taken from physical science where $X =$ electron as electric charge; $P + Q =$ electron as particle and wave; $A + B + C =$ electron as particle in the three aspects of electron, positron, and neutrino.

In each case, we have a transition from complementarity to a completely new situation in which the terms, though individually the same, have acquired quite new characters from their relationship to the others. This new systemic attribute is akin to **relatedness** but it is a dynamic relatedness that can be understood only if it is associated with notions of will and freedom. I have already discussed fully the connection among relatedness, will, freedom, and the triad in *The Dramatic Universe* and shall not attempt here to cover the same ground.[9] I must emphasize, however, that our understanding is immensely enriched when we take into account the systemic attribute of relatedness.

Whereas neither universality nor complementarity allow for connection among systems, relatedness establishes a nexus of connections that extends through all possible worlds. These many connections are possible because a term A of a triad X can also be a term of another triad Y, thus linking X and Y together. X can also be a term in a superordinate system Z. Thus, triadic relatedness can comprise coordination, subordination, and superordination. For example, we have A as husband and father in system X but son in system Y of the preceding generation. In system Y, A fulfills the role of system *C*, the child in *X*. The family *(ABC)* = *X* is a term in the system *Z* consisting of the three generations of grandparents *D*, parents *E*, and children *F*. In this way, *X* fulfills the same role as the link between *D* and *F*, as the child *C* fulfills as the link between *A* and *B*.

Evidently, the network of triads can be extended in all directions of space, time, and number. It can also be shown that any set of relations, however complex, can be reduced to a nexus of triads. It follows that relatedness is the systemic attribute of the triad, and, conversely, all cases of relatedness can be expressed as systems of the third order—that is, as triads.

THE TETRAD

If we contemplate universality, complementarity, and relatedness, we realize that we have not exhausted the attributes either of the world or of our own human nature. There is a certain rigidity in the network of triads that must be relaxed if we are to find place for creative activity. Moreover, no principle of order and, hence, continuity can be derived from the first three systems. We move, therefore, to the fourth-order system of the **tetrad.**[10]

We can approach the systemic attribute of the tetrad by recalling that freedom is the quintessence of relatedness. Freedom, to be realized, must be exercised, and we can ask the question "How and in what medium is freedom exercised?" The answer must surely be that the exercise of freedom is creative activity, and its medium is Being.[11] Creativity is the dynamic aspect of Being, and reciprocal to order, which is the passive aspect of Being. We can reasonably conclude that reciprocity is the systemic attribute of the tetrad, providing one understands that this reciprocity is not capricious, arbitrary, or transcendental, but the regulated, orderly activity whereby the world undergoes progressive enrichment of its content and quality. Herein lies the reciprocity of the tetrad.

The wealth of possible connections between four independent terms *K*, *L*, *M*, and *N* (there are twenty-four primary arrangements) entitles us to suppose that reciprocity has many forms. The entire process of existence in space and time is a form of creativity. The act of freedom whereby entirely new factors enter the process is another form. There is causal and there is non-causal reciprocity. There is absolute and there is relative reciprocity. To understand all the forms, we need to recognize the characters common to the four terms of any and every tetrad. We can deduce these characters from our intuition of the nature of any creative activity. There must be:

First term K: A motive force or source of the action;

Second term L: A medium or field in which the action proceeds;

Third term M: A character that represents the state of the system;

Fourth term N: A character that corresponds to the new element introduced by the creative action.

Figuratively, one could speak of *K* and *L* as "above and below" and *M* and *N* as "within and without." Creativity in all its forms involves interplay of the four factors in such a way as to transform and to enrich the situation. In another sense, it can be said that the four factors *K*, *L*, *M*, and *N* define order, which needs the concepts of "extremes" and "betweens."

To illustrate the tetrad, I take scientific activity as an example. Scientific activity is directed toward knowing, doing, and understanding. It is evidently creative in character and should therefore best be exemplified as a tetrad. We can, in fact, recognize four independent factors.

K: Understanding as the source of creative activity and also its goal;

L: Insight into nature as the field of scientific work;

M: The whole body of scientific knowledge representing the state of the system;

N: Experimentation, observation, and technical progress as the new element introduced by the creative action.

All scientific work requires, though in varying proportions, all four elements. The scientist must first make contact with his or her material. Insight into what lies before the scientist enables him or her to bring to bear both knowledge and experimental skill. As new data are discovered, the scientist must reconcile them with existing theories. Here, he or she must formulate a hypothesis—a situation that requires a creative act of understanding. As the hypothesis is tested and verified, it gradually enters into humankind's total understanding of nature. There is, therefore, a flux and reflux of creative activity in which all the factors play a role.

The four elements correspond to four abilities or skills that the accomplished scientist displays:

Insight: the ability to recognize significant characteristics in the subject matter of the research; the feeling or flair for natural phenomena.

Experimental skill: instrumentation and the conduct of experiments; power of observation.

Theoretical ability: analysis of results; empirical generalization; knowledge and memory.

Synthetic understanding: hypothesis formation; devising of crucial tests; integration of new ideas into existing body of theory and practice; true creativity in science.

A thorough examination of the tetrad reveals the immense significance of the systemic requirements of independence and mutual relevance. The "scientific tetrad" is applicable to the work of the individual, to team work, to the advancement of a particular branch of science, and to the scientific activity of humankind as a whole. This far-reaching generality of the systemic attribute of reciprocity illustrates the power of the tetrad as an instrument of understanding.

THE PENTAD

The first four systems of monad, dyad, triad, and tetrad by no means exhaust the modes of understanding open to us. The tetrad is the field of creativity, but it does not provide for non-creativity. It comprises all that becomes real but it does not allow for non-realization. Because of this lack, it has no central point from which alternative paths can bifurcate. We find this central point in the five-term system of the **pentad**.

In *The Dramatic Universe*, I connected the pentad with potentiality and the quality that I called "spiritualization."[12] We can also look upon the pentad as the focus of the creative agent. Referring back to the tetradic presentation of scientific activity, we might say that the tetrad needs to be completed by installing in the center the scientist himself or herself. Another way to consider the transition from tetrad to pentad is to notice that whereas the triad is too rigid, the tetrad is too lifeless to give an adequate representation of reality. To bring the tetrad to life, we must go forward and add a fifth term.

In the first volume of *The Dramatic Universe*, I showed how potential energy can be represented by adding the fifth dimension of eternity to the four dimensions of space-time.[13] Theoretical physics has not been able to formulate a theory of potential energy fields without adding an indepen-

dent parameter to the four coordinates of space-time. Looking more closely into the meaning of potentiality, we can see that it is the field of creativity, just as creativity is the field of freedom. To express the systemic attribute of the pentad, therefore, we must find a word that will convey the
notion of open potentiality within which creative action can be accomplished. We can then attach this word to that of the creative agent. The
word "spirituality" conveys some part of what we require, but it would
certainly be confusing to the reader who was not aware of the way we have
reached it. It seems better, therefore, to keep the word **potentiality**, making
clear that this term is to be taken subjectively as referring to the agent as
well as objectively as applied to the field.

Potentiality thus generalized is a notion of far-reaching significance. It
is closely associated with the idea of life itself, since life is both the creative agent and the field of creative action. In psychology, the fifth term is
the "I" or self that exercises the powers or functions of the psyche—the
latter, according to Carl Jung, consisting of four independent factors: sensation, intuition, thought, and feeling.

To assess the importance of understanding the pentad, we must remember that a system is a set of terms significantly connected. There are
very many ways in which five terms can be interconnected—a fact that
suggests that potentiality is a richer notion than is commonly supposed.
We tend to see it in terms of temporal successiveness, but there are probably more important forms of potentiality that lie outside the fields of
sense perception and mental constructs. Much that is mysterious and un-
accountable—the immense field of non-causal phenomena—probably refers to five-term systems that we incorrectly interpret as dyads and so miss
their true meaning.

THE HEXAD

In moving to the six-term system of the **hexad**, we identify the field in
which potentiality is realized. This situation is clearly the act itself
whereby creativity is accomplished. The additional term of the hexad
gives concreteness and uniqueness to the creative act. The nature of the
hexad is to provide the conditions for free and independent self-
realization. This situation can also be regarded as a complete event standing out from the undifferentiated goings-on of the existing world.

In *The Dramatic Universe*, I associated the hexad with recurrence and the sixth dimension that I called **hyparxis**.[14] This concept introduces a new depth and wealth of meaning, since it allows that which already is what is to become what it is. Self-realization in this sense does not mean transformation into something different, but *to become* in actuality what one was only in potentiality. There is a connection between these notions and the Thomist doctrine of *actus* whereby the world becomes real. Further, there are the notions of the self-contained field and of the completed Being who cannot only create but can do so within a world that is wholly his or her own.

For reasons that cannot be fully discussed here, I propose to use the word **significance** to designate the systemic attribute of the hexad.[15] We sometimes make the mistake of supposing that abstractions like "things," "ideas," or "people" can be significant. Significance can be ascribed only to the concrete event that stands out from the general stream of happenings. Even an idea can be at the heart of an event. "Universal suffrage" is an idea, which only became significant in the context of the reform that was an event. Without events, neither people nor ideas can rightly be called either significant or insignificant. Only events—or as Whitehead called them, actual occasions—can be said to exist concretely.[16] I must, however, sound a note of warning. To deserve the name, an event is not matter in motion within a limited region of space and time. It has a pattern and it has something more than that which, in *The Dramatic Universe*, I called "ableness-to-be."[17] The event asserts itself and reverberates through time and space. As the event occurs, it gains in concreteness. Starting as *potentia*, it becomes *actus*. In passing from the pentad to the hexad, the event acquires just that **significance** that I have taken to be the systemic attribute.

THE HEPTAD

A potential event can become actual only if there is a certain correspondence between its own character and the character of the historical environment. Failing this correspondence, an event—no matter how remarkable in its own right—fails to play its part in history. Such abortive events frequently occur at all scales, and their occurrence is evidence of the real distinction between hexad and the seven-term system of the **hep-

tad. The charge of the Light Brigade at Balaclava reverberates as an event, but it failed to become part of history. Indeed, the same can be said of the Crimean War as a whole. The death of Nicholas II deprived the event of its final act, and with the Treaty of Paris, history slipped back to the status quo of 1848. The distinction between abortive and integrative events throws much light upon the nature of the heptad, but we must know more of the character of the seven terms if we are to find the right word to express the systemic attribute.

The number seven has had a special, even sacred, character for both the Aryan and the Semitic cultures. Seven predominates in the Vedic and Avestan mythology and in the Hebrew and other Semitic traditions and rituals. The belief that there are seven primary qualities that attach to every important group of entities or powers is almost universal—seven colors, seven metals, seven planets, seven Pleiades, seven Maruts, seven virtues, seven sins, and—very significantly—seven tones in the musical scale.

Throughout all the septenaries there are also notions of an inner structure: primary and secondary colors, half- and full-tone intervals, major and minor planets, beneficial and harmful metals, and so forth. One intriguing study of the septenary is Alice Bailey's *Treatise on the Seven Rays*.[18] Here, there is a clear statement that the qualities of the septenary correspond to the seven main types of historical activities divided into three primary and four secondary characters. A different approach is that of Gurdjieff, who combines the notion of seven qualities with that of sevenfold progression from the event *in posse* to the event *in actu* as characteristic of all true history.[19]

Another important feature of the heptad is its connection with structure. Many traditional cultures believed that every harmonious structure has seven coordinated segments and that the human body can be presented as a septenary. These representations often depict the proportions and arrangement required to achieve harmony. In the same way, art and literature have often represented the human being in terms of a sevenfold nature. No doubt some fantasy and misunderstanding mar these schemes and the closely allied notion of microcosm and macrocosm linked by a common sevenfold structure. Nevertheless, behind these conceptions there may be genuine insights that could be brought into focus if we had the key to the systemic attribute of the heptad.

One way to search for this attribute is to see if there are any notions that cannot be expressed adequately with less than seven independent terms. Mechanical science demonstrates that a completely stable structure requires seven independent supports. Our preceding analysis indicates that the six-step sequence of universality, complementarity, relatedness, reciprocity, potentiality, and significance takes us to a concrete situation where we can describe events in all the wealth of form, function, creativity, freedom, and self-sufficing completeness that we find in our experience.

If we think of this sequence as a hexad, it gives us creativity realized in potentiality, but there remains another indispensable step—to connect a multitude of separate acts into a structure that is more significant, more concrete—in short, nearer to reality than the primitive diversity of mere happenings lacking in direction or meaning. Perhaps the best label to describe this step is **integration**, which suggests an operation both *within* the system to complete its own structure and *without* the system, which is thus brought into harmony with its environment.[20] The integrated system is an "integral" part of the entire historical realization by which existence itself acquires essential qualities that are not exemplified in the lower systems.

THE OCTAD

This possibility of an inwardly and outwardly integrated structure suggests that the progression of systems continues beyond seven terms. There are attributes that we can understand only in limited, partial, or specialized instances. We have intuitions of a realization that is deeper than history whereby the finite event acquires a limitless significance. We move, therefore to the eight-term system, or **octad**, which I represent as **individuality**. In *The Dramatic Universe*—for reasons more intuitive than rational—I take individuality to be the eighth category of fact.[21]

Associated with self-determination, individuality is beyond history and even beyond the integration of the heptad. Here, the integrated self becomes a source of free initiative—a creator in its own right. Individuality is the systemic attribute that initiates a fresh cycle of realization in which there is full cooperation between part and whole or, in the human situation, between human beings and God. In no simpler way could we reconcile the Infinite Omnipotence of God with the reality of human freedom and responsibility. Most discussions of this problem fail because they remain within the relatively abstract systems of the monad, dyad, and triad.

FINAL COMMENTS

I shall not carry the present discussion further, and the interested reader should refer to my full discussion of systems up to twelve in *The Dramatic Universe*.[22] My purpose in this essay has been to introduce systematics as a fundamental discipline of thought and action. I believe deeply that systematics provides a key to an informed understanding of many problems that, at present, issue in contradiction and confusion. Systematics can be understood as the complete development and generalization of the doctrine of the "principle of formal purposiveness," first proposed by Kant in *Critique of Judgement*.[23] This principle means that there is something to be understood and, further, that understanding itself is capable of unlimited progress because there is no limit to the series of **systemic attributes**, each of which penetrates more deeply into reality than its predecessor.[24] The idea of the correspondence between systems in nature and in thought is well expressed by Cassirer: "We find that nature 'favors' the effort of our faculty of judgment to discover a systematic order among her separate forms and, so to speak, meets it halfway."[25]

Table A describes the first eight systems, their attributes, and some of their terms. I believe strongly that once we grasp the notion of systems and see that every system constitutes a legitimate way of seeing and understanding the world, then spontaneously our awareness develops more deeply as we reflect upon every kind of situation that can arise in our experience. Systematics is not a science, if by that word we mean the study of a group of natural phenomena. Instead, systematics is an instrument applicable to all questions and problems. It is nearer to mathematics than to any other discipline. Indeed, it may be possible to show that mathematics is that branch of general systematics that deals solely with one-, two-, and three-term systems. Universality, complementarity, and relatedness are more than likely all the *a priori* notions required for deriving all the postulates and operations of mathematics.

<table>
<tr><td colspan="3">Table A
SYSTEMS, ATTRIBUTES, AND TERM CHARACTERS
FOR THE FIRST EIGHT SYSTEMS</td></tr>
<tr><td>SYSTEM</td><td>ATTRIBUTE</td><td>TERM CHARACTER</td></tr>
<tr><td>monad</td><td>universality</td><td>unity in diversity</td></tr>
<tr><td>dyad</td><td>complementarity</td><td>positive and negative</td></tr>
<tr><td>triad</td><td>relatedness</td><td>affirmation, receptivity, reconciliation</td></tr>
<tr><td>tetrad</td><td>reciprocity</td><td>unity, quantity, quality, diversity</td></tr>
<tr><td>pentad</td><td>potentiality</td><td>inner, higher, centric, lower, outer</td></tr>
<tr><td>hexad</td><td>significance*</td><td>self-realization</td></tr>
<tr><td>heptad</td><td>integration</td><td>completedness</td></tr>
<tr><td>octad</td><td>individuality</td><td>transcendence</td></tr>
</table>

*See note 15. —Ed.

If systems are constituents of reality and, if research and experiment help us to understand systems better, then systematics will prove to be an instrument of great importance for testing the validity of theories, for discovering hitherto unsuspected regularities in nature, and for helping us better to order our affairs, both personal and societal.

Never static, systematics involves an inherent dynamism that leads on from system to system and yet leaves nothing behind. As a system takes shape for our understanding and begins to disclose its systemic attribute, it also compels us to look beyond itself to a fuller, richer, and more concrete expression of the reality that we all share. There must be a dynamic, self-creative preparation for further transformations, each moment of which is only seemingly motionless. Whatever seeks fixation in static being is condemned to disintegration and nothingness. The dynamism of reality pervades all. As the poet Goethe explained:[26]

> *It has to move, to be creating deed,*
>
> *first make its form, then, changing it, proceed:*
>
> *All stopping, short—illusion's twist,*
>
> *For the Eternal onward moves in all,*
>
> *And into nothing everything must fall,*
>
> *If it in being would persist.*

NOTES

1. This article originally appeared in *Systematics*, vol. 1 (1963), pp. 5-18. It is reprinted with the permission of Elizabeth Bennett.

2. This condition distinguishes systems as here understood from those of Rudolf Carnap's *The Nature and Application of Inductive Logic* (Chicago: University of Chicago Press, 1951). The recognition of a term can be made with the help of elementary predicates, but the point is that each term of a system differs in respect of some elementary predicate from all the rest. See H. B. Curry, *Outlines of a Formalist Philosophy of Mathematics* (Amsterdam: North-Holland Publishers, 1958), p. 28.

3. On societies as systems, see Bennett, *The Dramatic Universe*, vol. 3, chap. 41.

4. David Bohm, *Causality and Chance in Modern Physics* (London: Routledge and Kegan Paul, 1961), pp. 104-128.

5. Bennett, *The Dramatic Universe*, vol. 1, pp. 49-54. This and following page references to volume I refer to the unabridged edition of 1956—Ed.

6. F. Bacon, *Advancement in Learning*, Book IV, art. 5 and 6 (Chicago: Encyclopedia Britannica, 1952).

7. For example, Jean Piaget, *Origins of Intelligence in Children* (New York: Norton, 1963).

8. The monad here is merely a one-term system and as such no more "real" than two-, three-, or multi-term systems. Nevertheless, it is worth noting that the monadology of Leibnitz is distinguished from atomism precisely by the property of monads of being universal. For Leibnitz, "each monad contains the whole infinity of existence within itself, and is thus a concentrated universe." See G. W. Leibnitz, *Monadology and Other Philosophical Writings* (New York: Oxford University Press, 1948), p. 706.

9. Bennett, *The Dramatic Universe*, Vol. 1, chaps. 27-31.

10. Bertrand Russell demonstrated that mathematical order can be defined only by reference to four independent terms. His conclusion agrees with the view here that the triad is not capable of supporting a principle of order. See Bertrand Russell, *Principles of Mathematics* (New York: Norton, 1938), pp. 364-373.

11. In volume 2 of *The Dramatic Universe*, I associated the tetrad with Being and

Creativity, but neither with order nor continuity, except in so far as it became clear that the notion of Being itself must be understood relatively and thus implies both order and continuity. See *The Dramatic Universe*, vol. 2, chaps. 32-34.

12. Bennett, *The Dramatic Universe*, vol. 2, pp. 295-297.

13. Bennett, *The Dramatic Universe*, vol. 1, pp. 156-163.

14. Bennett, *The Dramatic Universe*, vol. 1, pp. 166-170.

15. Note that in chapter 5, Bennett uses **significance** as a characteristic of the pentad. In volume 3 of *The Dramatic Universe*, he uses the term **coalescence** to describe the systemic attribute of the hexad (p. 48) and **significance** to describe the systemic attribute of the pentad (p. 39)—Ed.

16. A. N. Whitehead, *Process and Reality* (New York: MacMillan, 1960).

17. Bennett, *The Dramatic Universe*, vol. 1, pp. 132-135.

18. Alice Bailey, *Treatise on the Seven Rays* (New York: Lucis Publishers, 1936).

19. G. I. Gurdjieff, *All and Everything: Beelzebub's Tales to His Grandson* (New York: Harcourt Brace, 1950).

20. If we take coordination as the juxtaposition of two triads and, therefore, a hexad, integration is the binding together of two acts of will and, hence, a heptad.

21. Bennett, *The Dramatic Universe*, vol. 1, p 34, pp. 45-46.

22. Bennett, *The Dramatic Universe*, vol. 3, chap. 37.

23. I. Kant, *Critique of Judgment, Werke*, vol. 5, Introduction (Oxford: Clarendon Press, 1952), pp. 24-27.

24. Ibid. "The judgment has in itself a principle *a priori* of the possibility of nature... which... it assumes on behalf of a natural order cognizable by our understanding" (p. 26).

25. E. Cassirer, *The Problem of Knowledge* (New Haven, Connecticut: Yale University Press, 1950), p. 126.

26. These lines are from Goethe's poem, "One and All" ["*Eins und Alles*"] (1821). Both the English translation and the original German are in C. Middleton, Ed., *Johann Wolfgang von Goethe: Selected Poems* (Boston: Suhrkamp/Insel Publishers, 1983), pp. 240-243. The original German is:

> *Es soll sich regen, schaffend handeln,*
> *Erst sich gestalten, dann verwandeln;*
> *Nur scheinbar stehts Momente still.*
> *Das Ewige regt sich fort in allem:*
> *Denn alles muß in Nichts zerfallen,*
> *Wenn es im Sein beharren will.*

Further Reading

THE FOLLOWING bibliography is broken into two parts: writings by Bennett relating to systematics; and other relevant work. Presently, there are few published systematic studies. The seminal discussion is Bennett's four-volume *The Dramatic Universe*, published between 1956 and 1966. The most comprehensive presentation of systematics is chapter 37, "The Structure of the World," in volume 3. Reprints of these volumes can be ordered through Bennett Books, P.O. Box 1553, Santa Fe, New Mexico 87504 (505 989-8381).

From June 1963 to March 1974, Bennett and his research group, The Institute for the Comparative Study of History, Philosophy and the Sciences, published *Systematics*, a scholarly journal devoted largely to systematic studies. Some articles from this journal are listed below, but the interested reader should examine all eleven volumes. In 1987, to continue the development and application of systematics, Saul Kuchinsky, with assistance from the late Michael Franklin, established the Unis Institute, which, from 1987 to the present, has published *Unis: The Journal for Discovering Universal Qualities*. Information about the Institute and courses applying systematic principles can be obtained by writing the Unis Institute, Box 6615, Bridgewater, New Jersey 08807.

WORKS BY J. G. BENNETT RELEVANT TO SYSTEMATICS

Creative Thinking. Charles Town, West Virginia: Claymont Communications, 1989.

Dramatic Universe, The. Vol. 1, *The Foundations of Natural Philosophy.* Abridged by Eric Mandel. Charles Town: Claymont Communications, 1987.

Dramatic Universe, The. Vol. 2, *The Foundations of Moral Philosophy.* Charles Town: Claymont Communications, 1987.

Dramatic Universe, The. Vol. 3, *Man and His Nature.* Charles Town: Claymont Communications, 1987.

Dramatic Universe, The. Vol. 4, *History.* Charles Town: Claymont Communications, 1987.

Enneagram Studies. York Beach, Maine: Samuel Weiser, 1983.

General Systematics, *Systematics,* vol. I, no. 1 (1963), pp. 5-18 [reprinted in *Unis,* vol. I (1987), pp. 9-27, and in appendix A of this book].

Gurdjieff: Making a New World. Santa Fe, N.M.: Bennett Books, 1992.

Judgment and Strategy in Management Communication Analysis [mimeographed article]. London: Structural Communications Systems, Ltd. [no date].

Progress and Hazard, *Systematics,* vol. 5 (1968), pp. 319-330.

A Spiritual Psychology. Coombe Springs: Coombe Springs Press, 1964.

Witness. Charles Town, West Virginia: Claymont Communications, 1983.

OTHER WORKS RELEVANT TO SYSTEMATICS

Bortoft, Henri. Counterfeit and Authentic Wholes: Finding a Means for Dwelling in Nature, pp. 281-302 in David Seamon and Robert Mugerauer, eds., *Dwelling, Place and Environment: Towards a Phenomenology of Person and World.* New York: Columbia University Press, 1989.

Bortoft, Henri. *Goethe's Scientific Consciousness.* Tunbridge Wells, England: The Institute for Cultural Research, 1986; Hudson, New York: Lindesfarne Press, 1993.

Bortoft, Henri. The Whole: Counterfeit and Authentic, *Systematics,* vol. 9 (1971), pp. 43-73.

Hodgson, Anthony M. Birth to Adulthood: A Systematic Study, *Systematics,* vol. 3 (1963), pp. 230-269.

Hodgson, Anthony M. Educational Curricula: Their Design, Testing and Application, *Systematics,* vol. 1 (1963), pp. 157-172.

King, Clarence E. Systematics of a Manufacturing Process, *Systematics,* vol. 1 (1963), pp. 111- 126.

Kuchinsky, Saul. *Systematics: Search for Miraculous Management.* Charles Town, West Virginia: Claymont Communications, 1986.

Low, Albert. *Zen and Creative Management.* New York: Doubleday, 1976.

Pledge, K. W. Structured Process in Scientific Experiments, *Systematics,* vol. 3 (1963), pp. 304-333 (reprinted in J. G. Bennett, *Enneagram Studies* [see above], pp. 84-122).

Shantock Systematics Group. *A Systematics Handbook.* Sherborne, Glouchestershire: Coombe Springs Press, 1975.

[no author]. Systematics and General Systems Theory, *Systematics,* vol. 1 (1963), pp. 105-109.

[no author]. Systematics and Systems Theories, *Systematics,* vol. 7 (1970), pp. 273-278.

[no author]. Systematics in Operation, *Systematics,* vol. 1 (1963), pp. 185-187.

GENERAL BIBLIOGRAPHY OF J.G. BENNETT

Concerning Subud. London: Hodder and Stoughton, 1960.

The Crisis in Human Affairs. London: Hodder and Stoughton, 1954.

Creative Thinking. Charles Town, WV: Claymont Communications, 1989.

Dramatic Universe, The. Vol. 1, *The Foundations of Natural Philosophy.* Abridged by Eric Mandel. Charles Town: Claymont Communications, 1987.

Dramatic Universe, The. Vol. 2, *The Foundations of Moral Philosophy.* Charles Town: Claymont Communications, 1987.

Dramatic Universe, The. Vol. 3, *Man and His Nature.* Charles Town: Claymont Communications, 1987.

Dramatic Universe, The. Vol. 4, *History.* Charles Town: Claymont Communications, 1987.

Deeper Man. London: Turnstone Books, 1985.

Enneagram Studies. York Beach, Maine: Weiser, 1983.

Energies—Material, Vital, Cosmic. Charles Town: Claymont Communications, 1989.

Gurdjieff: A Very Great Enigma. York Beach: Weiser, 1983.

Gurdjieff: Making a New World. Santa Fe, NM: Bennett Books, 1992.

How We Do Things. Charles Town: Claymont Communications, 1989.

Hazard: The Risk of Realization. Santa Fe, NM: Bennett Books, 1991.

Is There "Life" on Earth? An Introduction to Gurdjieff. Santa Fe: Bennett Books, 1989.

Idiots in Paris (with E. Bennett). York Beach, ME: Weiser, 1991.

Intimations: Talks with J. G. Bennett at Beshara. New York: Weiser, 1975.

Journeys to Islamic Countries. Vols. 1 & 2. Sherborne: Coombe Springs Press, 1977.

Long Pilgrimage: The Life and Teaching of Sri Govindananda Bharati, Known as the Shivapuri Baba. California: Dawn Horse Press, 1983.

Masters of Wisdom. London: Turnstone Press, 1980.

Needs of a New Age Community: Talks on Spiritual Community and Fourth Way Schools. Santa Fe: Bennett Books, 1990.

Sacred Influences: Spiritual Action in Human Life. Santa Fe: Bennett Books, 1989.

Sevenfold Work, The. Charles Town: Claymont Communications, 1979.

Sex. York Beach: Weiser, 1989.

A Spiritual Psychology. Lakemont, Georgia: CSA Press, 1974.

The Spiritual Hunger of the Modern Child (with others). Charles Town: Claymont Communications, 1984.

Sufi Spiritual Techniques. Ellingstring, England: Coombe Springs Press, 1982.

Talks on Beelzebub's Tales. York Beach: Weiser, 1988.

Transformation. Charles Town: Claymont Communications, 1978.

Way to Be Free, The. New York: Weiser, 1980.

What Are We Living For? Santa Fe, NM: Bennett Books, 1991.

Witness: The Autobiography of John G. Bennett. Charles Town: Claymont Communications, 1983.

○

<u>Bennett Books' Publications</u>
by J.G.Bennett

in print
Sacred Influences: Spiritual Action in Human Life, 1989
Is There "Life" on Earth? An Introduction to Gurdjieff, 1989
Needs of a New Age Community: Talks on
Spiritual Community and Fourth Way Schools, 1990
Hazard: The Risk of Realization, 1991
What Are We Living For?, 1991
Gurdjieff: Making a New World, 1992
Elementary Systematics, 1993

forthcoming
Deeper Man, 1993
Man's Task and His Reward, 1993
Existence, 1993

BENNETT BOOKS
P.O. Box 1553
Santa Fe, N.M. 87504
505 989-8381 / 986-1428

Books that Support Spiritual Development
Write or call for a complete catalog

Thank You for Purchasing This Book

Use this page to order directly from Bennett Books and receive free shipping *and* a 20% discount when your order is $25 or more.

All titles are by J.G. Bennett

Quantity / Title (paper, unless noted)	**Price**	**Extended**
__Elementary Systematics	$12.95	_______
__Gurdjieff: Making a New World	$17.95	_______
__What Are We Living For?	$11.00	_______
__Hazard: The Risk of Realization	$11.00	_______
__Hazard: The Risk of Realization (cloth)	$22.00	_______
__Hazard (cloth + paper)	$31.00	_______
__Is There "Life" on Earth?	$9.50	_______
__Needs of a New Age Community	$11.00	_______
__Sacred Influences	$8.00	_______
__The Planetary Enneagram	$2.75	_______
__Enneagram Studies	$7.95	_______
__Gurdjieff: A Very Great Enigma	$5.95	_______
__Sex	$6.95	_______
__Talks on Beelzebub's Tales	$8.95	_______
__Idiots in Paris	$9.95	_______
__Dramatic Universe, Volume 2	$16.95	_______
__Dramatic Universe, Volume 3	$16.95	_______
__Dramatic Universe, Volume 4	$19.95	_______
__Energies	$11.95	_______
__How We Do Things	$9.95	_______
__Creative Thinking	$10.95	_______

Sub Total _______

if subtotal is *less* than $25 *add* $4.50 shipping   _______

if subtotal is $25 or *more* deduct 20% −   _______

N.M. residents please add current sales tax   _______

Total enclosed   _______

Check or MO in $U.S. drawn on a U.S. bank only; payable to:
Bennett Books
P.O. Box 1553, Santa Fe, N.M. 87504
For information call 505 989-8381 / 986-1428

Name____

Address____

City______________________State_____ ZIP_________

Phone____________________ Date___________________

EMS92